ENERGY FUNCTION ANALYSIS FOR POWER SYSTEM STABILITY

THE KLUWER INTERNATIONAL SERIES IN ENGINEERING AND COMPUTER SCIENCE

POWER ELECTRONICS AND POWER SYSTEMS

Consulting Editor

Thomas A. Lipo
University of Wisconsin - Madison

Other books in the series:

SPOT PRICING OF ELECTRICITY
Fred C. Schweppe
ISBN 0-89838-260-2

RELIABILITY ASSESSMENT OF LARGE ELECTRIC POWER SYSTEMS
Roy Billinton and Ronald N. Allan
ISBN 0-89838-266-1

MODERN POWER SYSTEMS CONTROL AND OPERATION
Atif S. Debs
ISBN 0-89838-265-3

ELECTROMAGNETIC MODELLING OF POWER ELECTRONIC CONVERTERS
J. A. Ferreira
ISBN 0-7923-9034-2

ENERGY FUNCTION ANALYSIS FOR POWER SYSTEM STABILITY
M. A. Pai
ISBN 0-7923-9035-0

ENERGY FUNCTION ANALYSIS FOR POWER SYSTEM STABILITY

by

M. A. Pai
University of Illinois at Champaign/ Urbana

KLUWER ACADEMIC PUBLISHERS
Boston/Dordrecht/London

Distributors for North America:
Kluwer Academic Publishers
101 Philip Drive
Assinippi Park
Norwell, Massachusetts 02061 USA

Distributors for all other countries:
Kluwer Academic Publishers Group
Distribution Centre
Post Office Box 322
3300 AH Dordrecht, THE NETHERLANDS

Library of Congress Cataloging-in-Publication Data

Pai, M. A., 1931-
Energy function analysis for power system stability / by M.A. Pai.
p. cm. — (The Kluwer international series in engineering and computer science. Power electronics and power systems)
Bibliography: p.
Includes index.
ISBN 0-7923-9035-0
1. Electric power systems—Mathematical models. I. Title.
II. Series: Kluwer international series in engineering and computer science. Power electronics & power systems.
TK1005.P329 1989
621.31—dc20 89-15387
CIP

Printed in the United States of America.

Dedicated to

Jawaharlal Nehru

CONTENTS

PREFACE

This research monograph is in some sense a sequel to the author's earlier one (Power System Stability, North Holland, New York 1981) which devoted considerable attention to Lyapunov stability theory, construction of Lyapunov functions and vector Lyapunov functions as applied to power systems. This field of research has rapidly grown since 1981 and the more general concept of energy function has found wide spread application in power systems. There have been advances in five distinct areas (i) Developing energy functions for structure preserving models which can incorporate non-linear load models (ii) Energy functions to include detailed model of the generating unit i.e., the synchronous machine and the excitation system (iii) Reduced order energy functions for large scale power systems, the simplest being the single machine infinite bus system (iv) Characterization of the stability boundary of the post-fault stable equilibrium point (v) Applications for large power networks as a tool for dynamic security assessment. It was therefore felt appropriate to capture the essential features of these advances and put them in a somewhat cohesive framework. The chapters in the book roughly follow this sequence. It is interesting to note how different research groups come to the same conclusion via different reasonings. For example the concept of critical and non-critical groups of machines may be looked at from the vulnerable cutset point of view, lowest normalized potential energy margin, slow coherency concept or simply based on acceleration, kinetic energy etc. of the machine at $t = 0^+$ or $t = t_{c\ell}$. Unfortunately the terminology is not standard in the literature and hence there has been some difficulty in explaining in great detail some of the more recent research work. However the book will provide a basis for the researcher, theoretical or practically oriented one to explore the topic in greater detail. The last chapter lists some of the topics which merit further investigation.

I would like to thank all the researchers whose work is explained in the book. I know I might have left out some other important work also, to whom I apologize. This is the dilema an author faces in writing a book where the literature is extensive and the terminology is not standard. Useful discussions with Professor P. W. Sauer at University of Illinois, Professor K. R. Padiyar at I.I.Sc., Bangalore, India, Professor V. Vittal at Iowa State University, and Professor H. D. Chiang at Cornell University are gratefuly acknowledged.

I would like to thank my wife Nandini for the moral and logistic support during the course of the last two years when the book was being written. I would like to thank Kelly C. Voyles of our publications office for her excellent typing of the manuscript and Fred Daab for the fine job of doing the drawings.

I would like to acknowledge the support of the National Science Foundation for supporting my research in this area through its grant ECS 84-14677 and ECS 87-19055.

The book is dedicated to Mr. Jawaharlal Nehru, the first Prime Minister of India and whose birth Centenary (1889-1989) is being celebrated this year. His vision about Science and Technology being the cornerstone for India's development has made many of us knowingly or unknowingly grateful to him.

M. A. Pai

ENERGY FUNCTION ANALYSIS FOR POWER SYSTEM STABILITY

Chapter 1

POWER SYSTEM STABILITY IN SINGLE MACHINE SYSTEM

1.1 Introduction

An interconnected power system consists of generating units (including turbine-governor and excitation systems), transmission lines, loads, transformers (including off-nominal ratio and phase shifting transformers), static var compensators and HVDC lines. The size of the interconnection varies but the technical problems are the same. It is not uncommon for some utilities to study systems which have 10,000 buses. As a rule of thumb about one fourth of these buses will have generators connected. At the planning level, after examining the adequacy of transmission capacity, economy of generation etc., the planner would invariably study the stability of the system for a set of disturbances ranging from the worst, such as three phase to ground fault (whose probability of occurrence is rare) to single phase faults which constitute about 70% of the disturbances. There are good production grade programs which simulate the time behavior of the system for such disturbances. There are literally thousands of variables that can be monitored even on a moderate sized system such as a 500 bus system. The planner wants to see if a fault is properly cleared without the system losing synchronism. A system is said to be synchronously stable (i.e., retain synchronism) for a given fault if the system variables settle down to some steady state values as $t \to \infty$. These simulation studies are called transient stability studies. A typical transient stability study with detailed modeling for a 500 bus 100 machine system might take up to an hour. Hence the planner limits himself to a few likely scenarios of fault occurrences.

In actual operation, however, the system loading conditions and parameters may be quite different from those assumed at the planning stage. Thus to ensure power system security against possible abnormal conditions due to contingencies, the operator would like to simulate the contingencies in advance, assess the results, and then take preventive control action if necessary. This whole process is called dynamic security assessment (DSA) and preventive control. A similar approach already exists for static security assessment and preventive control and modern energy control centers have software to carry out these functions. However, very little progress has been made in regard to DSA and preventive control. The reason is simply that it takes a long time to conduct a transient simulation even for a single contingency. Even with simplified models of the machines the operator has to look at the plots of rotor angles and velocities of several machines to make a judgement about stability. It is nice to have a direct way of making a stabilty/instability decision by computing a single number such as a transient stability index (TSI). TSI can indicate whether you are stable or unstable and, if stable, what is the margin of stability. In its simplest form, TSI is the difference between the critical clearing time of a circuit breaker to isolate the disturbance and the actual time for which it is set. To make it more meaningful, one could convert this information into a dimensionless normalized stability margin or, even better, convert the information in terms of line flow limits, vulnerable transmission lines, etc.

There are basically two approaches for computing the TSI

1) Use super-fast time simulation techniques for the selected set of contingencies and convert the totality of the information into operator oriented guide lines.

2) Use direct method such as the energy function or the Lyapunov method to compute the critical energy which the disturbed system can absorb before the fault is cleared and the system is restored to a stable state.

With the present state-of-the-art in computer technology, the first method is not feasible at the moment although research has begun in the area. The early pioneers in using the energy function method were Magnusson (1947), Aylett (1958), Gless (1966) and El-Abiad and Nagappan (1966). While Aylett (1958) proposed an energy integral approach, El-Abiad and Nagappan (1966) put the method in the context of Lyapunov's stability theory. After 1966 there has been a significant effort in this area and these are well documented in the survey papers of Fouad (1975), Ribbens-Pavella and Evans (1985), Varaiya, Wu and Chen (1987) and the book by Pai (1981). Since 1981 there has been a tremendous amount of literature in the stability analysis of power systems using direct methods, both theoretical and practical. This book is mainly aimed at covering that material. It is pertinent to point out that until 1978, the method remained largely conservative until Kakimoto et al. (1978) and Athay et al. (1979) suggested alternate but closely related approaches to compute the critical energy dependent on the particular disturbance. These are the potential energy surface (PEBS) method and the relevant or controlling u.e.p. (unstable equilibrium point) method, respectively. The next breakthrough came in terms of accommodating more detailed modeling because of the structure preserving approach proposed by Bergen and Hill (1981). Mathematical characterization of the stability region which explains both the method of Kakimoto et al. (1978) and Athay et al. (1979) is to be found in the paper by Chiang et al. (1987). On the application side there has been extensive testing of the method of Athay et al. (1979) with modifications by

Fouad et al. (1987) on large scale realistic systems. Thus, one can confidently argue that the method, after nearly four decades of research, holds promise for DSA. Preventive control methodologies are in an early research stage with the work of Pai et al. (1984) and El-Kady et al. (1986) based on sensitivity concepts. The preceding brief summary gives a historical perspective on the problem.

1.2 Statement of the Stability Problem

The definitions that follow defining stability of a power system are based on the IEEE Committee Report (1982) and CIGRE Committee 32 Report (1978). There are two kinds of disturbances defined; small and big disturbances. A small disturbance is one for which the dynamics of the power system can be linearized. A big disturbance is one where such linearization is not valid and hence the mathematical model is a set of nonlinear differential equations with or without algebraic constraints.

The question of stability is whether for a given disturbance, large or small, the trajectories of pre-disturbance operating quantities of the system during the disturbance remain in the domain of attraction of the post-disturbance equilibrium when the disturbance is removed. This concept is one of transient stability. Obviously, transient stability is the property of the power system, its steady-state operating condition <u>and</u> the disturbance itself. Since the disturbance may be large, protective relaying may change the power system configuration and the post-disturbance steady-state operating condition, if reached, may be quite different from the pre-disturbance steady-state operating condition. In some extreme cases the post disturbance steady state may be reached after losses of many lines, loads and generators and it may not be an acceptable mode of operation. Thus a power system can be considered transiently stable only if it reaches an acceptable steady state. The loss of synchronism in a purely AC system is the usual indication of instability. However, in a more complex system involving DC lines or asynchronous operation, loss of synchronism by slipping of a pole by a machine may or may not be an acceptable state depending on the circumstances and the operating criteria. Thus we come to the following mathematical definition.

Transient Stability implies that an acceptable post-disturbance steady-state operating condition of the power system is asymptotically stable and the response to the given disturbance is such that the trajectories of the operating quantities tend to this operating condition as time increases.

In this book we confine to purely AC systems so that the loss of synchronism is equivalent to instability. We define synchronous operation more precisely following the IEEE Committee report (1982).

Synchronous operation of a machine

A machine is in synchronous operation with a network or another machine to which it is connected if its average electrical speed (product of its rotor angular velocity and the number of pole pairs) is equal to the angular frequency of the AC network voltage or to the electrical speed of the other machine.

Synchronous operation of a power system

A power system is in synchronous operation if all its connected synchronous machines are in synchronous operation with the AC network and with each other.

From the preceding discussion it is clear that if a machine tends to go unstable by the loss of synchronism, its angular velocity with respect to other machines goes on increasing. This is also true for a group of machines which go unstable with respect to the rest of the system. The rotor angle δ_i of each machine is measured with respect to a fixed reference frame. Hence, instability of a machine means that the rotor angle of machine i pulls away from the rest of the system. Thus relative rotor angles rather than absolute rotor angles need to be monitored to test stability/instability. Figure 1.1 shows the rotor angles for the case of stability and instability.

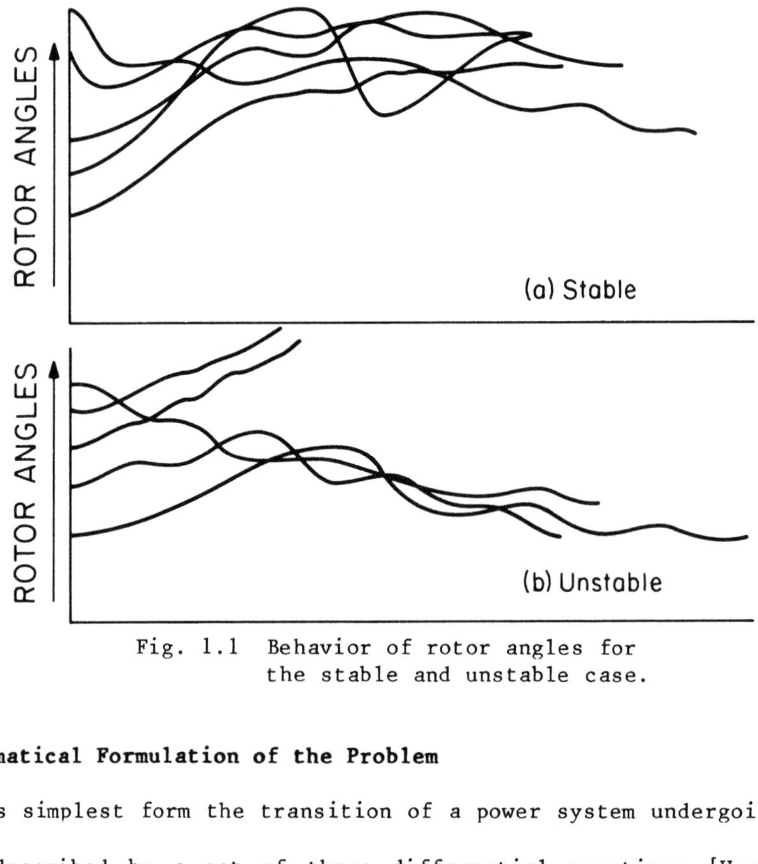

Fig. 1.1 Behavior of rotor angles for the stable and unstable case.

1.3 Mathematical Formulation of the Problem

In its simplest form the transition of a power system undergoing a disturbance is described by a set of three differential equations [Varaiya et al. (1987)]

$$\dot{x}(t) = f^{I}\left(x(t)\right) \qquad -\infty < t \leq 0 \tag{1.1}$$

$$\dot{x}(t) = f^{F}\left(x(t)\right) \qquad 0 < t \leq t_{c\ell} \tag{1.2}$$

$$\dot{x}(t) = f\left(x(t)\right) \qquad t_{c\ell} < t < \infty \tag{1.3}$$

$x(t)$ is the vector of state variables of the system at time t. At $t = 0$ a fault occurs and the dynamics change from f^I to f^F. During $0 < t \leq t_{c\ell}$, called the faulted period, the system is governed by the fault-on dynamics f^F. Actually before the fault is cleared at $t = t_{c\ell}$ we may have several switchings in the network each giving rise to a different f^F. For simplicity we have taken a single f^F indicating that there are no structural changes between $t = 0$ and $t = t_{c\ell}$. When the fault is cleared at $t = t_{c\ell}$, we have the post-fault dynamics $f\left(x(t)\right)$. In the prefault state $-\infty < t \leq 0$, the system would have settled down to a steady state so that $x(0) = x_0$ is known. We, therefore, need not discuss (1.1). We are then left with

$$\dot{x}(t) = f^F\left(x(t)\right) \qquad 0 < t \leq t_{c\ell} \tag{1.4}$$

$$x(0) = x_0$$

and

$$\dot{x}(t) = f\left(x(t)\right) \qquad t > t_{c\ell} \tag{1.5}$$

with the initial condition $x(t_{c\ell})$ for (1.5) provided by the solution of (1.4) evaluated at $t = t_{c\ell}$. Viewed in another manner, the solution of (1.4) provides at each instant of time the possible initial conditions for (1.5). Let us assume that (1.5) has a stable equilibrium point x_s. The question is whether the trajectory $x(t)$ for (1.5) with the initial condition $x(t_{c\ell})$ will converge to x_s as $t \to \infty$. The largest value of $t_{c\ell}$ showing this to be true is called the critical clearing time t_{cr}.

From the above discussion it is clear that if we have an accurate estimate of the region of attraction of x_s, then t_{cr} is obtained when the trajectory of (1.4) exits the region of attraction (Fig. 1.2). The computation of the region

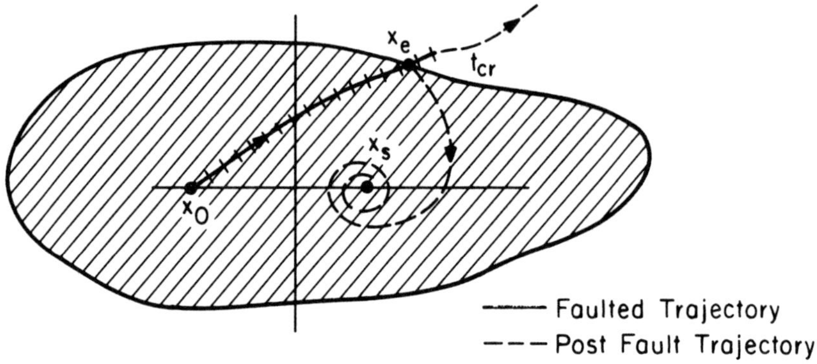

Fig. 1.2 Region of attraction and computation of t_{cr}.

of attraction for a general nonlinear dynamical system is far from easy. In the case of power systems with simple machine models, the characterization of this region has been discussed by Chiang et al. (1987). However, for practical application one has to resort to some approximate methods. The most popular among these methods is to describe the interior of the region of attraction through an inequality of the type $V(x) < V_{cr}$ where $V(x)$ is the Lyapunov or energy function for (1.5). The computation of V_{cr} is a difficult step and there are currently three methods with a number of variations on each method. In these methods it is assumed that the region of attraction is formed by connecting the unstable equilibrium points (u.e.p.'s) surrounding x_s through particular multi-dimensional surfaces called manifolds.

These methods are

1) $V_{cr} = V(x^{uc})$ where x^{uc} is the unstable equilibrium point (u.e.p.) resulting in the lowest value of V_{cr} among the u.e.p.'s lying on the stability boundary of (1.5).

2) $V_{cr} = V(x^{ur})$ where x^{ur} is the relevant or controlling u.e.p., i.e., the u.e.p. closest to the point where the fault-on trajectory of (1.4) exits

the region of attraction of (1.5).

3) V_{cr} = Maximum value of the potential energy component of V(x) along the trajectory of (1.4).

We will be discussing these methods in detail in subsequent chapters.

The computation of t_{cr} therefore involves the following steps.

1. Computing x_s knowing the post-disturbance structure
2. Formulating V(x) for (1.5)
3. Computation of V_{cr}. This step will involve integrating (1.4) for a short interval in method (2) and for a longer time in method (3). There is no integration involved in method (1) for computing V_{cr}. Method (1) is known to give very conservative results while methods (2) and (3) are comparable.
4. Calculating the time instant t_{cr} when $V(x) = V_{cr}$ on the faulted trajectory of (1.4). The faulted trajectory needs to be integrated for all the three methods to obtain t_{cr}. In method (3), the faulted trajectory is already available while computing V_{cr}.

1.4 Modeling Issues

The above steps assume that we have the model

$$\dot{x} = f^F\left(x(t)\right) \tag{1.6}$$

and

$$\dot{x} = f\left(x(t)\right) \tag{1.7}$$

In reality the model is a set of differential-algebraic equations, i.e.,

$$\dot{x} = f^F\left(x(t),\ y(t)\right) \tag{1.8}$$

$$0 = g^F\left(x(t),\ y(t)\right) \tag{1.9}$$

and

$$\dot{x} = f\left(x(t),\ y(t)\right) \tag{1.10}$$

$$0 = g\left(x(t),\ y(t)\right) \tag{1.11}$$

The function g represents the nonlinear algebraic equations governing the network and loads while the differential equations represent the dynamics of the generating unit and its controls. In power system literature, the modeling of equations in the form (1.8) and (1.9) or (1.10) and (1.11) has been covered extensively. Various reduced order models are also discussed in the literature, the simplest of them leading to an elimination of the algebraic equations, resulting in either (1.6) or (1.7). Models that involve algebraic equations can incorporate nonlinear load models and lead to the concept of structure preserving energy functions while models consisting of differential equations only lead to closed form types of energy functions in certain cases.

It is not at all clear at this stage as to whether a more detailed model will lead to more accurate estimates of t_{cr}. What appears to be true, however, from extensive simulation studies by researchers is that for the so called first swing stability (i.e., instability occurring in less than 1 sec.) the classical model and loads represented as constant impedances resulting in equations of the type (1.6) and (1.7) will suffice. Most of the research prior to 1981 using direct methods uses this and it is referred in the literature as the internal node model. For stability over a larger time interval (several seconds) detailed models may be justified.

1.5 Motivation Through a Single Machine Infinite Bus (SMIB) System

In spite of the extensive research in the area, it appears from the literature that when a system goes unstable it separates into two parts with one group of machines, called the "critical" machines, separating from the rest of the system. Hence, application oriented reseach has centered around identifying these critical machines and replacing them with a single equivalent machine, with the rest of the system being replaced by another equivalent machine. We then have a two-machine system which in turn can be replaced by a single-machine infinite bus (SMIB) system. Conceptually, this key concept can be found in the pioneering work of Kimbark (1948, pg. 111), which also enabled the equal area criterion to be formulated. We discuss the two-machine case which will be useful in later chapters as a basic concept.

We consider the following three cases:

1. Combination of two finite machines to form a SMIB system.
2. Equal area criterion and energy function for the SMIB system.
3. Energy function for a two-machine system.

1.5.1. Combination of Two Machines (Kimbark [1948])

Let the swing equations of the two machines be

$$M_1 \frac{d^2\delta_1}{dt^2} = P_{m1} - P_{e1} \tag{1.12}$$

$$M_2 \frac{d^2\delta_2}{dt^2} = P_{m2} - P_{e2} \tag{1.13}$$

where P_{mi} (i=1, 2) are the mechanical inputs and P_{ei} (i=1, 2) are the electrical outputs of the machines. Let the relative rotor angle be denoted by $\delta = \delta_1 - \delta_2$. Then

$$\frac{d^2\delta}{dt^2} = \frac{d^2\delta_1}{dt^2} - \frac{d^2\delta_2}{dt^2} = \frac{(P_{m1} - P_{e1})}{M_1} - \frac{(P_{m2} - P_{e2})}{M_2} \tag{1.14}$$

Multiplying both sides of (1.14) by M_1M_2/M_1+M_2

$$\frac{M_1 M_2}{M_1+M_2}\frac{d^2\delta}{dt^2} = \frac{1}{(M_1 + M_2)}\left[M_2 \left(P_{m1} - P_{e1}\right) - M_1 \left(P_{m2} - P_{e2}\right)\right]$$

$$= \frac{1}{(M_1 + M_2)}\left[M_2 P_{m1} - M_1 P_{m2} - \left(M_2 P_{e1} - M_1 P_{e2}\right)\right]$$

i.e.,

$$M_{eq}\frac{d^2\delta}{dt^2} = P_m - P_e \tag{1.15}$$

where

$$P_m = \frac{1}{M_1 + M_2}\left(M_2 P_{m1} - M_1 P_{m2}\right)$$

$$P_e = \frac{1}{M_1 + M_2}\left(M_2 P_{e1} - M_1 P_{e2}\right) \tag{1.16}$$

$$M_{eq} = M_1 M_2 / M_1 + M_2$$

Consider a two-machine system (Fig. 1.3). After eliminating the passive nodes

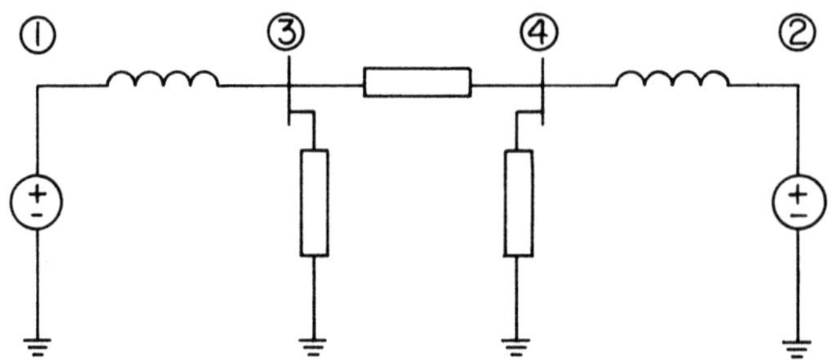

Fig. 1.3 Two-machine system.

(3) and (4) by network reduction, let the representation at nodes (1) and (2) be

$$Y = \begin{bmatrix} Y_{11} & Y_{12} \\ Y_{21} & Y_{22} \end{bmatrix} \tag{1.17}$$

If $Y_{ij} = |Y_{ij}| \angle\theta_{ij}$, then the electrical powers P_{e1} and P_{e2} can be derived as

$$P_{e1} = |E_1|^2 \, |Y_{11}| \cos \theta_{11} + |E_1||E_2||Y_{12}| \cos(\theta_{12} - (\delta_1 - \delta_2))$$

$$= |E_1|^2 \, G_{11} + |E_1||E_2||Y_{12}| \cos(\theta_{12} - (\delta_1 - \delta_2))$$

$$P_{e2} = |E_2|^2 \, G_{22} + |E_1||E_2||Y_{12}| \cos(\theta_{21} - (\delta_2 - \delta_1)) \tag{1.18}$$

Since Y is symmetric $Y_{12} = Y_{21}$ and hence $\theta_{12} = \theta_{21}$. Substituting these values in (1.16) and letting $\delta = \delta_1 - \delta_2$ we get

$$P_e = \frac{M_2 |E_1|^2 G_{11} - M_1 |E_2|^2 G_{22}}{M_1 + M_2} + \frac{|E_1||E_2||Y_{12}| \, [M_2 \cos(\delta-\theta_{12}) - M_1 \cos(\delta+\theta_{12})]}{M_1 + M_2} \tag{1.19}$$

The two cosine terms involving δ can be combined into a single cosine term to simplify P_e as (algebra omitted)

$$P_e = P'_e + P_e^{max} \sin (\delta-\gamma) \tag{1.20}$$

where

$$P'_e = \frac{M_2 \, |E_1|^2 G_{11} - M_1 \, |E_2|^2 G_{22}}{M_1 + M_2} \tag{1.21}$$

$$P_e^{max} = \frac{|E_1||E_2||Y_{12}| \sqrt{M_1^2 + M_2^2 - 2M_1M_2 \cos 2\theta_{12}}}{M_1 + M_2} \tag{1.22}$$

$$\gamma = - \tan^{-1} \left(\frac{M_1 + M_2}{M_1 - M_2} \tan \theta_{12}\right) - 90° \tag{1.23}$$

Special Cases

(i) If $M_2 = \infty$, we get the SMIB case, i.e., (1.15) becomes

$$M_1 \frac{d^2\delta}{dt^2} = P_{m1} - \left(|E_1|^2 G_{11} + P_e^{max} \sin (\delta-\gamma)\right) \tag{1.24}$$

where $P_e^{max} = |E_1|\ |E_2|\ |Y_{12}|$, $\gamma = \theta_{12} - 90°$.

(ii) If the network is purely reactive, then $G_{11} = G_{22} = 0$ and $\theta_{12} = \pi/2$ so that $P_e = |E_1||E_2||Y_{12}| \sin\delta$ from (1.19). One of the machines (say 2) acts like a motor and the other (machine 1) is a generator. This means that $P_{m1} = -P_{m2} = P_m$. The swing equation (1.15) becomes

$$\frac{M_1 M_2}{M_1+M_2}\ \frac{d^2\delta}{dt^2} = P_m - |E_1||E_2||Y_{12}| \sin\delta \tag{1.25}$$

$$M_{eq}\frac{d^2\delta}{dt^2} = P_m - P_e^{max} \sin\delta$$

The power angle curve is independent of machine inertias. We next consider the case of single machine infinite bus case (1.24) with $\theta_{12} = 90°$, which is mathematically equivalent to (1.25).

1.5.2. Energy Function and Equal Area Criterion for SMIB System (No Transfer Conductances, i.e. $\theta_{12} = 90°$)

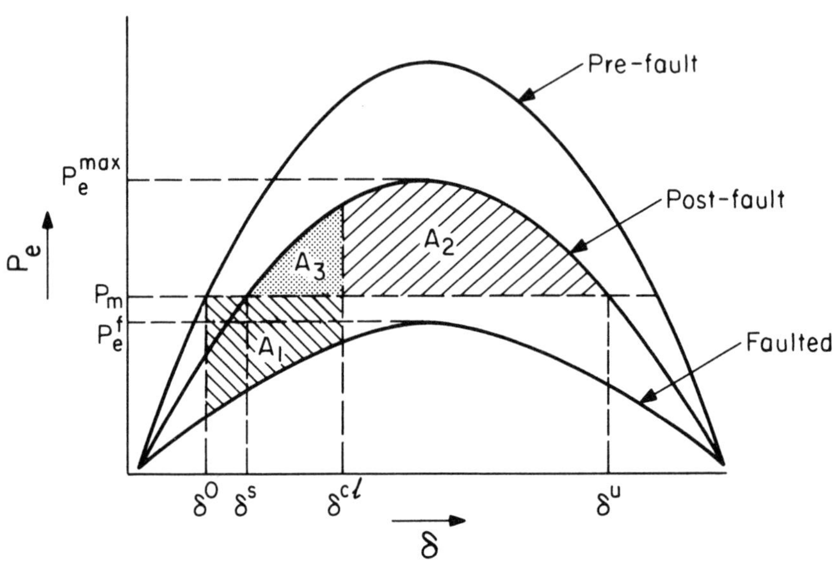

Fig. 1.4 Equal area criterion for SMIB case.

Let the post-fault system be described by the swing equation ($G_{11} = 0$)

$$M \frac{d^2\delta}{dt^2} = P_m - P_e^{max} \sin \delta \qquad (1.26)$$

The post-fault stable equilibrium point is given by $\delta^s = \sin^{-1} \left(\frac{P_m}{P_e^{max}}\right)$ and the nearest unstable equilibrium point given by $\delta^u = \Pi - \delta^s$. The right-hand side of (1.26) can be expressed as the negative gradient of a potential energy function V_{PE}, i.e.,

$$M \frac{d^2\delta}{dt^2} = - \frac{\partial V_{PE}(\delta)}{\partial \delta} \qquad (1.27)$$

where $V_{PE} = - P_m \delta - P_e^{max} \cos \delta$.

Multiplying (1.27) by $\frac{d\delta}{dt}$ on both sides and integrating we get

$$\frac{d}{dt} \left(\frac{1}{2} M \left(\frac{d\delta}{dt}\right)^2 + V_{PE}(\delta)\right) = 0 \qquad (1.28)$$

Since $\frac{d\delta}{dt} = \omega$, this implies $\frac{dV}{dt} = 0$ where

$$V(\delta.\omega) = \frac{1}{2} M \omega^2 + V_{PE}(\delta) + C , \qquad (1.29)$$

with C as a constant of integration. The constant C is adjusted so that $V(\delta^s,0) = 0$. Therefore, $C = - V_{PE}(\delta^s)$. Hence, the energy function (1.29) is given by

$$V(\delta,\omega) = \frac{1}{2} M \omega^2 + V_{PE}(\delta) - V_{PE}(\delta^s)$$

$$= \frac{1}{2} M \omega^2 - P_m(\delta-\delta^s) - P_e^{max} (\cos\delta - \cos\delta^s) \qquad (1.30)$$

The region of attraction of $(\delta^s,0)$ is defined by the inequality $V(\delta.\omega) < V_{cr}$. In the case of a single machine system, V_{cr} is determined as $V_{cr} = V(\delta^u,0)$, i.e.,

$$V_{cr} = -P_m(\delta^u - \delta^s) - P_e^{max}(\cos\delta^u - \cos\delta^s) \tag{1.31}$$

Since $\delta^u = \Pi - \delta^s$, we get V_{cr} as

$$V_{cr} = -P_m(\Pi - 2\delta^s) + 2P_e^{max}\cos\delta^s \tag{1.32}$$

The system is stable for $t = t_{c\ell}$ if along the faulted trajectory $V(\delta,\omega) < V_{cr}$ at $t = t_{c\ell}$ and the critical clearing t_{cr} is obtained when $V(\delta,\omega) = V_{cr}$ on the faulted trajectory.

Now we shall show that the energy function results are identical to the equal area criterion. From Fig. 1.4 the system is stable for a given $\delta_{c\ell}$ (obtained by integrating the faulted trajectory up to $t_{c\ell}$) if $A_1 < A_2$ by the equal area criterion. Physically A_2 is the ability of the post-disturbance system to absorb energy (potential) and A_1 is the kinetic energy injected into the system during the fault. If we add area A_3, then the stability condition becomes (letting P_e^f = maximum value of the faulted power angle curve)

$$A_1 + A_3 < A_2 + A_3 \tag{1.33}$$

Now

$$\begin{aligned} A_1 &= \int_{\delta^o}^{\delta^{c\ell}} (P_m - P_e^f \sin\delta)\, d\delta \\ &= \int_{\delta^o}^{\delta^{c\ell}} (P_m - P_e^f \sin\delta)\, d\delta \\ &= \int_{\delta^o}^{\delta^{c\ell}} M\frac{d\omega}{dt}\, d\delta \\ &= \int_{\delta^o}^{\delta^{c\ell}} M\frac{d\omega}{dt}\,\omega dt = \int_{o}^{\omega^{c\ell}} M\,\omega\, d\omega \\ &= \frac{1}{2}M(\omega^{c\ell})^2 \end{aligned} \tag{1.34}$$

$$A_3 = \int_{\delta^s}^{\delta^{c\ell}} (P_e^{max} \sin \delta - P_m)\, d\delta = - P_e^{max} \cos \delta^{c\ell} + P_e^{max} \cos \delta^s$$

$$- P_m (\delta^{c\ell} - \delta^s) \tag{1.35}$$

Changing $\delta^{c\ell}$, $\omega^{c\ell}$ to any $\delta \cdot \omega \cdot$ we get

$$A_1 + A_3 = \frac{1}{2} M\omega^2 - P_m (\delta-\delta^s) - P_e^{max} (\cos \delta - \cos \delta^s)$$

$$= V(\delta,\omega) \tag{1.36}$$

Also it can be shown that

$$A_2 + A_3 = \int_{\delta^s}^{\Pi-\delta^s} (P_e^{max} \sin \delta - P_m)\, d\delta$$

$$= 2 P_e^{max} \cos \delta^s - P_m (\Pi-2\delta^s)$$

$$= V_{cr} \tag{1.37}$$

Equations (1.36) and (1.37) are the same as (1.30) and (1.32), respectively. Thus we have shown that the energy function method is equivalent to the equal area criterion in the case of the SMIB system.

1.5.3. Energy Function for Two-Machine System

It has been shown in Sec. 1.5.1 that the two-machine system can be converted into an equivalent single-machine system of the form

$$M_{eq} \frac{d^2\delta}{dt^2} = P_m - \left(P_e' + P_e^{max} \sin(\delta-\gamma)\right) \tag{1.38}$$

Define $M_{eq} = M$, $P_m - P_e' = P_m'$ and $\delta-\gamma = \delta'$. We then have the SMIB system of the form

$$M \frac{d^2\delta'}{dt^2} = P_m' - P_e^{max} \sin \delta' \tag{1.39}$$

The theory of the preceding section can be applied. Alternatively the first integral for Eq. (1.38) can be constructed following a procedure similar to the no transfer conductance case in Sec. 1.5.2. Letting the parameters of Eq. (1.38) refer to the post-fault system, we have

$$M_{eq} \frac{d^2\delta}{dt^2} = -\frac{\partial V_{PE}(\delta)}{\partial \delta} \tag{1.40}$$

where $V_{PE}(\delta) = -\left(P_m - P'_e\right)\delta - P_e^{max}\cos(\delta-\gamma)$. As before, letting $V_{PE}(\delta^s) = 0$, we get

$$V(\delta,\omega) = \frac{1}{2} M_{eq}\,\omega^2 - \left(P_m - P'_e\right)(\delta-\delta^s) - P_e^{max}\left[\cos(\delta-\gamma) - \cos(\delta^s-\gamma)\right] \tag{1.41}$$

V_{cr} is obtained by letting $\omega = 0$ and $\delta = \delta^u = \Pi - \delta^s + 2\gamma$. To calculate δ^u we can also use Eq. (1.39) as follows. The stable equilibrium point of (1.39) is $\hat{\delta}^s = \sin^{-1}\left(\frac{P'_m}{P_e^{max}}\right)$. Therefore, $\delta^s = \hat{\delta}^s + \gamma$. Also, $\hat{\delta}^u = \Pi - \hat{\delta}^s$. It follows $\delta^u = \hat{\delta}^u + \gamma = \Pi - \hat{\delta}^s + \gamma = \Pi - \delta^s + 2\gamma$ One can also show this via the power angle curves. Since γ may be different for the pre-fault, faulted and post-fault periods, the respective power angle curves are displaced sinusoids and the equal area criterion can also be derived [Xue et al. (1988)]. With this motivation for energy functions for single- and two-machine systems, we can gain a better insight into multimachine systems which we consider in subsequent chapters. The outlines of rest of the chapters is given below.

1.6 Chapter Outline

In Chapter 2 we develop the internal node model of a power system based on the classical model of the synchronous machine and treating loads as constant impedances. Then we develop a system wide energy function for this as well as the structure preserving model. Both nonlinear voltage dependent and frequency dependent loads are considered in the structure preserving model.

Chapter 3 is devoted to reduced order energy functions such as the individual machine energy function, cutset energy function and the time scale energy function.

In Chapter 4 we consider energy functions for detailed models based on the internal node as well as the structure preserving descriptions.

Chapter 5 discusses the characterization of the region of stability and some of the theoretically based techniques to compute it.

Chapter 6 discusses the use of the energy function for computing critical clearing time for realistic, large scale systems and also its use as a tool for dynamic security assessment.

Chapter 7 discusses open research problems. Appendix A gives the data for the widely used 10 machine 39 bus system.

1.7 Conclusion

This chapter has introduced the basic physical notions of synchronous stability and has developed a functional framework for analyzing stability. The concept of energy function is illustrated through a single-machine, infinite bus system.

Chapter 2

ENERGY FUNCTIONS FOR CLASSICAL MODELS

2.1 Introduction

The bulk of literature in stability analysis of power systems using the direct method employs the so-called classical model for the generator. The advantage of this model is that the mechanical rotor angle deviation with respect to the synchronous reference is directly related to the electrical equivalent circuit known as the voltage behind the transient reactance model. To be precise, the phase angle of this voltage is the rotor angle deviation with respect to the synchronous frame of reference. For decades, power engineers have found this model to be adequate to predict the stability/instability of the system over a time interval of 0-1 sec, also known as the "first-swing" stability. Analytically this model, in a multi-machine framework, is easier to handle. In this chapter we first develop energy functions using this model for multi-machine systems using the internal node representation. This is possible only if we treat the loads as constant impedances. Next, using the classical model and preserving the structure of the network we develop the system energy function in which it is possible to include non-linear loads.

A detailed exposition of system wide energy function using internal node representation is contained in Pai (1981). Hence, we will abbreviate this part and devote more attention to the energy functions using the structure preserving model.

2.2 Internal Node Representation

2.2.1 Mathematical model

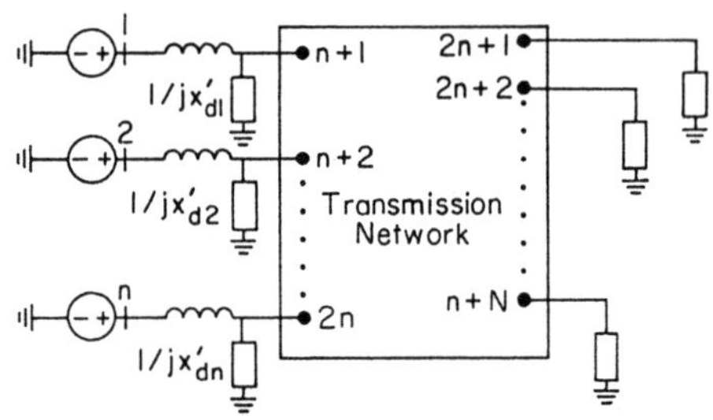

Figure 2.1 Multi-machine power system.

In Figure 2.1 the transmission network has N nodes numbered n+1 ... 2n, 2n+1, ... n+N, of which the last N-n are purely load nodes and the first n are generator nodes where there may also be loads. The loads are assumed to be constant impedances and computed in the form of admittances as

$$y_{\ell i} = \frac{P_{\ell i} - jQ_{\ell i}}{|V_i|^2}, \quad i = n+1, \ldots n+N \tag{2.1}$$

where the load is $P_{\ell i} + jQ_{\ell i}$ at the ith bus. The transmission network has the nodal admittance matrix Y_{BUS} as

$$Y_{BUS} = \begin{array}{c} \begin{matrix} \;\; n & \quad N-n \end{matrix} \\ \begin{bmatrix} Y_1 & Y_2 \\ Y_3 & Y_4 \end{bmatrix} \end{array} \begin{matrix} n \\ N-n \end{matrix} \tag{2.2}$$

We now (i) augment this Y_{BUS} matrix by admittances corresponding to the generator transient reactances as

$$y_i = (1/jx'_{di}), \quad i = 1, 2 \ldots n \tag{2.3}$$

to create the internal nodes 1, 2 ... n.

(ii) Add the admittances $y_{\ell i}$ corresponding to the loads at the buses (n+1)...n+N. This will result in an overall network admittance matrix as

$$\hat{Y}_{BUS} = \begin{array}{c} \\ \end{array}\overset{\begin{array}{ccc} n & n & N-n \end{array}}{\begin{bmatrix} y & -y & 0 \\ -y & y+Y_1+y_{\ell g} & Y_2 \\ 0 & Y_3 & Y_4+y_{\ell\ell} \end{bmatrix}} \tag{2.4}$$

where $y_{\ell g} = \text{Diag}\,(y_{\ell i}),\ i = n+1, \ldots 2n$

$y_{\ell\ell} = \text{Diag}\,(y_{\ell i}),\ i = 2n+1, \ldots n+N$

$y = \text{Diag}\,(1/jx'_{di}),\ i = 1, 2 \ldots n$

In the $\hat{Y}_{BUS}$ matrix only the first n internal buses are retained and the rest N are eliminated in order to get the internal node model. For this purpose, symbolically $\hat{Y}_{BUS}$ is partitioned as

$$\hat{Y}_{BUS} = \overset{\begin{array}{cc} n & N \end{array}}{\begin{bmatrix} Y_A & Y_B \\ Y_C & Y_D \end{bmatrix}} \begin{array}{c} n \\ N \end{array} \tag{2.5}$$

and the network equations for this augmented network are

$$\begin{array}{c} n \\ N \end{array} \begin{bmatrix} I_A \\ 0 \end{bmatrix} = \begin{bmatrix} Y_A & Y_B \\ Y_C & Y_D \end{bmatrix} \begin{bmatrix} E_A \\ V_D \end{bmatrix} \tag{2.6}$$

where E_A is the vector of voltages behind the transient reactances and V_D is the vector of physical bus voltages. Retaining the first n nodes (internal nodes) only gives

$$I_A = \left(Y_A - Y_B\, Y_D^{-1}\, Y_C\right) E_A$$
$$= Y_{int}\, E_A \tag{2.7}$$

Let the elements of Y_{int} be denoted as $Y_{ij} = G_{ij} + j\, B_{ij}$ (2.8)

The swing equations of the n machines are given by

$$\frac{H_i}{\pi f} \frac{d^2 \delta_i}{dt^2} + D_i \frac{d\delta_i}{dt} = P_{mi} - P_{gi}\ , \quad i = 1, 2 \ldots n \tag{2.9}$$

H_i = Inertia constant in seconds of the ith machine

D_i = Damping constant in sec/radian of the ith machine

P_{mi} = Mechanical input to the ith machine (assumed constant) in per unit

P_{gi} = Electrical power output of the ith machine in per unit

δ_i = Rotor angle of the ith machine with respect to a synchronously rotating reference frame in radians.

2.2.2 Analytical Expression for P_{gi}

$$P_{gi} = \text{Re}\left[E_i\, I_i^*\right]$$
$$= \text{Re}\left[E_i \sum_{j=1}^{n} Y_{ij}^*\, E_j^*\right] \tag{2.10}$$

Let $E_i = |E_i| \angle \delta_i$

$$P_{gi} = \text{Re}\left[\,|E_i| \sum_{j=1}^{n} \left(G_{ij} - j\, B_{ij}\right) |E_j| \angle \delta_i - \delta_j\right]$$
$$= \text{Re}\left[\sum_{j=1}^{n} |E_i|\,|E_j| \left(\cos\delta_{ij} + j\sin\delta_{ij}\right)\left(G_{ij} - j\, B_{ij}\right)\right]$$
$$= |E_i|^2\, G_{ii} + \sum_{\substack{j=1\\ \neq i}}^{n} \left(|E_i|\,|E_j|\, B_{ij} \sin\delta_{ij} + |E_i|\,|E_j|\, G_{ij}\cos\delta_{ij}\right) \tag{2.11}$$

where $\delta_{ij} = \delta_i - \delta_j$

Let $|E_i||E_j|\, B_{ij} = C_{ij}$, $|E_i||E_j|\, G_{ij} = D_{ij}$

Then

$$P_{gi} = |E_i|^2 G_{ii} + \sum_{\substack{j=1 \\ \neq i}}^{n} \left(C_{ij} \sin \delta_{ij} + D_{ij} \cos \delta_{ij}\right) \tag{2.12}$$

The swing equations are

$$\frac{H_i}{\pi f} \frac{d^2\delta_i}{dt^2} + D_i \frac{d\delta_i}{dt} = P_{mi} - P_{gi} \tag{2.13}$$

Let $\frac{H_i}{\pi f} = M_i$, $P_i = P_{mi} - |E_i|^2 G_{ii}$

Then
$$M_i \frac{d^2\delta_i}{dt^2} + D_i \frac{d\delta_i}{dt} = P_i - \sum_{\substack{j=1 \\ \neq i}}^{n} \left(C_{ij} \sin \delta_{ij} + D_{ij} \cos \delta_{ij}\right) \tag{2.14}$$

which can be written as

$$M_i \frac{d^2\delta_i}{dt^2} + D_i \frac{d\delta_i}{dt} = P_i - P_{ei} (\delta_1 \ldots \delta_n), \quad i=1, 2 \ldots n \tag{2.15}$$

Let α_i be the rotor angle with respect to a fixed reference. Then $\delta_i = \alpha_i - \omega_s t$. $\dot{\delta}_i = \frac{d\alpha_i}{dt} - \omega_s \triangleq \omega_i - \omega_s$, where ω_i is the angular velocity of the rotor and ω_s is the synchronous speed in radians per sec. Thus both δ_i and $\dot{\delta}_i$ are expressed with respect to a synchronously rotating reference frame. Equation (2.15) is converted to a set of first order differential equations by introducing the state variables δ_i and ω_i as

$$\begin{aligned} \dot{\delta}_i &= \omega_i - \omega_s \\ \dot{\omega}_i &= \frac{1}{M_i} \left(P_i - P_{ei} (\delta_1 \ldots \delta_n) - D_i (\omega_i - \omega_s)\right) \end{aligned} \tag{2.16}$$

If these equations pertain to the post disturbance system they are equivalent to Eq. (1.7), after transferring the equilibrium point of (2.16) to the origin namely,

$$\dot{x} = f\left(x(t)\right) \tag{2.17}$$

2.3 Energy Functions for Internal Node Models

Prior to 1979, there was considerable research in constructing a Lyapunov function for the system (2.16). However, analytical Lyapunov functions can only be constructed if the transfer conductances are zero, i.e., $D_{ij} \equiv 0$. Since these terms have to be accounted properly, first integrals of motion of the system are constructed and they are called energy functions. We have seen in Chapter I that the behavior of relative rotor angles is the key for judging stability/instablity. Hence, we have two options; either to use the relative rotor angle formulation or the center of inertia formulation.

2.3.1 Relative rotor angle formulation

Choose some machine, preferably the one with heavy inertia, as the reference machine and define rotor angles with respect to this machine. If the nth machine is the reference machine, Eq. (2.14) then becomes

$$\dot{\delta}_{in} = \omega_i - \omega_n \qquad i=1,\ 2,\ \ldots\ n-1$$

$$\dot{\omega}_i = M_i^{-1}\left\{-D_i\,(\omega_i-\omega_s) + P_i - \sum_{\substack{j=1\\ \neq i}}^{n} C_{ij}\sin(\delta_{in}-\delta_{jn}) + D_{ij}\cos(\delta_{in}-\delta_{jn})\right\}$$
$$i=1,\ 2,\ \ldots\ n \tag{2.18}$$

The energy function for this system with $D_i \equiv 0$ has been shown by Ribbens-Pavella and Evans (1985) to be the sum of a path independent integral $V_{KE}(\omega) + V_p(\delta)$ plus a path dependent integral $V_d(\delta)$ due to the terms D_{ij} terms as follows

$$V(\delta,\omega) = V_{KE}(\omega) + V_{PE}(\delta)$$
$$= V_{KE}(\omega) + V_p(\delta) + V_d(\delta) \tag{2.19}$$

where

$$V_{KE}(\omega) = \frac{1}{2M_T} \sum_{i=1}^{n-1} \sum_{j=i+1}^{n} M_i M_j \omega_{ij}^2 \tag{2.20}$$

$$V_p(\delta) = \sum_{i=1}^{n-1} \sum_{j=i+1}^{n} \{ -\frac{1}{M_T} (P_i M_j - P_j M_i)(\delta_{ij} - \delta_{ij}^s) \tag{2.21}$$
$$- \sum_{i=1}^{n} \left(C_{ij} (\cos \delta_{ij} - \cos \delta_{ij}^s)\right)\}$$

$$V_d(\delta) = \frac{1}{M_T} \int_{\delta_{ij}^s}^{\delta_{ij}} \sum_{i=1}^{n-1} \sum_{j=i+1}^{n} \left[M_j \left(\sum_{k \neq i}^{n} D_{ik} \cos \delta_{ik} \right) - M_i \left(\sum_{k \neq j}^{n} D_{jk} \cos \delta_{jk} \right) \right] d\delta_{ij} \tag{2.22}$$

δ^s is the vector of the post-fault stable equilibrium point and $M_T = \sum_{i=1}^{n} M_i$.

2.3.2 Center of inertia formulation

The second option is to use the angle of the center of inertia (COI) as the reference angle since it represents the "mean motion" of the system. Although the resulting energy function is identical to $V(\delta,\omega)$ it has the advantage of being more symmetric and easier to handle in terms of the path dependent terms. Modern literature invariably uses the COI formulation originally proposed by Tavora and Smith (1971). The energy function in the COI notation including D_{ij} terms (transfer conductances) was first proposed by Athay et al. [1979].

We derive the energy function for the conservative system (assuming $D_i \equiv 0$). Defining the center of inertia (COI) for the whole system as

$$\delta_o = \frac{1}{M_T}\sum_{i=1}^{n} M_i\delta_i \text{ and center of speed as } \omega_o = \frac{1}{M_T}\sum_{i=1}^{n} M_i\,\omega_i \tag{2.23}$$

where $M_T = \sum_{i=1}^{n} M_i$, we transform the variables δ_i, ω_i to the COI variables as $\theta_i = \delta_i - \delta_o$, $\tilde{\omega}_i = \omega_i - \omega_o$. The swing equations (2.14) with $D_i = 0$ become (omitting the algebra)

$$M_i \frac{d^2\theta_i}{dt^2} = P_i - \sum_{\substack{j=1\\ \neq i}}^{n} [C_{ij} \sin\theta_{ij} + D_{ij}\cos\theta_{ij}] - \frac{M_i}{M_T} P_{COI}$$

$$\triangleq f_i(\theta) \qquad i=1, 2 \ldots n \tag{2.24}$$

where $P_i = P_{mi} - |E_i|^2 G_{ii}$; $P_{COI} = \sum_{i=1}^{n} P_i - 2\sum_{i=1}^{n-1}\sum_{j=i+1}^{n} D_{ij}\cos\theta_{ij}$

Corresponding to the faulted and post-fault state we have two sets of differential equations, as:

$$M_i \frac{d\tilde{\omega}_i}{dt} = f_i^F(\theta)$$

$$\frac{d\theta_i}{dt} = \tilde{\omega}_i \qquad 0 < t < t_{cl} \qquad i=1, 2, \ldots n \tag{2.25}$$

and

$$M_i \frac{d\tilde{\omega}_i}{dt} = f_i(\theta)$$

$$\frac{d\theta_i}{dt} = \tilde{\omega}_i \qquad t \geq t_{cl} \qquad i=1, 2, \ldots n \tag{2.26}$$

Let the post-fault system have the stable equilibrium point at $\theta = \theta^s$, $\tilde{\omega} = 0$, which is obtained by solving the non-linear algebraic equations

$$f_i(\theta) = 0 \quad , \quad i=1, 2, \ldots n$$

Since $\sum_{i=1}^{n} M_i \theta_i = 0$, θ_n is expressed in terms of the other θ_i's and substituted in the above set of equations which is then equivalent to

$$f_i (\theta_1 \ldots \theta_{n-1}) = 0 \; , \; i=1, 2 \ldots n-1 \tag{2.27}$$

The basic procedure for computing the critical clearing time consists of the following steps:

1. Construct an energy or Lyapunov function $V(\theta,\tilde{\omega})$ for the system (2.26), i.e., the post-fault system.
2. Find the critical value of $V(\theta,\tilde{\omega})$ for a given fault denoted by V_{cr}.
3. Integrate the equations (2.25), i.e., the faulted equations until $V(\theta,\tilde{\omega}) = V_{cr}$. This instant of time is called the critical clearing time t_{cr}.

While this procedure is common to all the methods mentioned in Sec. 1.3, they differ from one another in steps 2 and 3, i.e., finding V_{cr} and integrating the swing equations. There is general agreement that the first integral of motion of (2.26) constitutes a good energy function. An alternative derivation to that of Athay et al. (1979) is presented below (Pai (1984)).

From (2.26) we have for i = 1, 2 ... n

$$dt = \frac{M_1 d\tilde{\omega}_1}{f_1(\theta)} = \frac{d\theta_1}{\tilde{\omega}_1} = \frac{M_2 d\tilde{\omega}_2}{f_2(\theta)} = \frac{d\theta_2}{\tilde{\omega}_2} = \ldots \frac{M_n d\tilde{\omega}_n}{f_n(\theta)} = \frac{d\theta_n}{\tilde{\omega}_n} \tag{2.28}$$

Integrating the pairs of equations for each machine between $(\theta_i^s, 0)$ to $(\theta_i, \tilde{\omega}_i)$ results in

$$V_i (\theta, \tilde{\omega}) = \frac{1}{2} M_i \tilde{\omega}_i^2 - \int_{\theta_i^s}^{\theta_i} f_i (\theta) \, d\theta_i \; , \quad i = 1, 2, \ldots n \tag{2.29}$$

Adding these for all the machines we get the first integral of motion for the system as (omitting the algebra)

$$V(\theta,\tilde{\omega}) = \frac{1}{2}\sum_{i=1}^{n} M_i \tilde{\omega}_i^2 - \sum_{i=1}^{n}\int_{\theta_i^s}^{\theta_i} f_i(\theta)\, d\theta_i \tag{2.30}$$

In the literature, expression (2.29) is referred to as the individual machine energy function (Michel et al. (1983)) and expression (2.30) as the system energy function.

Expansion of (2.29) and (2.30) yields respectively (omitting the algebra)

$$V_i(\theta,\tilde{\omega}) = \frac{1}{2} M_i \tilde{\omega}_i^2 - P_i(\theta_i - \theta_i^s) + \sum_{\substack{j=1\\ \neq i}}^{n} \Big(C_{ij}\int_{\theta_i^s}^{\theta_i} \sin\theta_{ij}\, d\theta_i$$

$$+ \sum_{\substack{j=1\\ \neq i}}^{n} D_{ij}\int_{\theta_i^s}^{\theta_i} \cos\theta_{ij}\, d\theta_i\Big) + \frac{M_i}{M_T}\int_{\theta_i^s}^{\theta_i} P_{COI}\, d\theta_i \tag{2.31}$$

and

$$V(\theta,\tilde{\omega}) = \sum_{i=1}^{n} V_i(\theta,\tilde{\omega})$$

$$= \frac{1}{2}\sum_{i=1}^{n} M_i \tilde{\omega}_i^2 - \sum_{i=1}^{n} P_i(\theta_i - \theta_i^s) - \sum_{i=1}^{n-1}\sum_{j=i+1}^{n} \Big(C_{ij}(\cos\theta_{ij} - \cos\theta_{ij}^s)$$

$$- \int_{\theta_i^s + \theta_j^s}^{\theta_i + \theta_j} D_{ij}\cos\theta_{ij}\, d(\theta_i + \theta_j)\Big) \tag{2.32}$$

since

$$\sum_{i=1}^{n} \frac{M_i}{M_T}\int_{\theta_i^s}^{\theta_i} P_{COI}\, d\theta_i = 0$$

Note that both (2.31) and (2.32) contain path dependent integrals. In view of this we cannot assert that V_i and V are positive definite. If $D_{ij} \equiv 0$, it has been shown by Willems (1971), Pai (1972) through a Luré-Popov formulation and use of the Moore-Anderson theorem that $V(\delta, \omega)$ is positive definite. Hence $V(\theta,$

$\tilde{\omega}$) is positive definite since it is equivalent to $V(\delta, \omega)$. For details see Pai (1981). With $D_{ij} \neq 0$, it has been observed through numerous simulations, that $V(\theta,\tilde{\omega})$ is constant and it can be shown analytically that $\dot{V}(\theta,\tilde{\omega}) = 0$. Moreover, when a fault occurs, one injects energy into the post-fault system which is always positive (Racz and Bokay (1988)).

2.4 Individual Machine and Other Energy Functions

The individual machine energy function (IMEF) (2.30) contains less terms than the system energy function but it still requires the integration for all the machines in the faulted state in order to get θ_j's ($j \neq i$). The concept of IMEF could be extended to groups of machines where each group consists of a set of coherent (ideally) but nearly-coherent (practically) machines. Each group is replaced by its COI and connections between the groups is replaced by a single equivalent connection. From the accelerations at $t = 0^+$ or from the fault-on trajectory, one can identify the group that separates from the rest of the system. Then an equivalent two machine can be formed which can be reduced to a single machine infinite bus system [Fouad and Vittal (1983), Lemmon et al. (1988), Xue et al. (1988), and Padiyar and Ghosh (1987)]. This kind of equivalencing assumes that if instability occurs the system separates into two groups. One can also interpret this as a cut set energy function by designating the "fictitious" lines between the separating group and the rest of the system as a vulnerable cut set [Chandrasekhar & Hill (1983)]. By staying at the group level and also separately retaining the dynamics of machines within each group, one gets what is popularly known in the literature as the slow-coherency based model [Chow (1982)]. It is possible to postulate a slow energy function between the COI's of the groups and the fast energies of each of the groups. These energies are known as time-scale energies [Pai et al. (1986)]. These various reduced order energy functions will be discussed in the next chapter.

2.5 Structure Preserving Energy Functions

2.5.1 Mathematical model

The assumption of a constant impedance load may introduce an error in t_{cr} computation when in fact the loads are nonlinear and voltage dependent. The first step in removing this deficiency in formulation was by Bergen and Hill (1981) who assumed frequency dependent load and constant voltage at the buses. Narasimhamurthy and Musavi (1984) considered constant active power and voltage dependent reactive power loads while Tsolas et al. (1985) considered P-Q type loads. Padiyar and Sastry (1987) have included nonlinear voltage dependent loads for both real and reactive powers. Van-Cutsem and Ribbens-Pavella (1985) have a general formulation which includes, in addition, the frequency dependent terms for the real power loads. We follow their formulation procedure below.

Let us first adopt the following notation

- n rotor angles $\delta_1 \ldots \delta_n$
- n relative rotor velocities $\dot{\delta}_1 \ldots \dot{\delta}_n$, $(\dot{\delta}_i = \omega_i - \omega_s)$
- N bus voltage magnitudes $V_{n+1} \ldots V_{n+N}$
- N bus phase angles $\psi_{n+1} \ldots \psi_{n+N}$

The vectors ω, δ, ψ are defined as

$$\omega = \begin{bmatrix} \omega_1 \\ \cdot \\ \cdot \\ \cdot \\ \omega_n \end{bmatrix} \qquad \delta = \begin{bmatrix} \delta_1 \\ \cdot \\ \cdot \\ \cdot \\ \delta_n \end{bmatrix} \qquad \psi = \begin{bmatrix} \psi_{n+1} \\ \cdot \\ \cdot \\ \cdot \\ \psi_{n+N} \end{bmatrix} \tag{2.33}$$

To have a consistent set of equations we re-define the overall phase angle vector as

$$\phi = [\delta^T, \psi^T]^T \tag{2.34}$$

and the overall voltage vector as

$$V = [V_1 \dots V_n \; V_{n+1} \dots V_{n+N}]^T \tag{2.35}$$

with $V_i = E_i$ (i = 1, 2 ... n). In the transmission network, the resistance of the transmission lines are ignored since they are small compared to the reactance. We then get a Y_{BUS} matrix including the internal nodes of order (N+n) whose elements are jB_{ij}. The structure of the matrix is similar to (2.5) but does not include the loads.

System Equations

The swing model for the ith machine is

$$M_i \ddot{\delta}_i = P_{mi} - P_{gi}(V, \phi) \tag{2.36}$$

These are expressed as a set of first order d.e's as

$$\begin{aligned} \dot{\phi}_i &= (\omega_i - \omega_s) \\ M_i \dot{\omega}_i &= P_{mi} - \sum_{j=1}^{n+N} B_{ij} V_i V_j \sin(\phi_i - \phi_j), \quad i = 1, 2 \dots n \end{aligned} \tag{2.37}$$

Though general, the summation term on the right-hand side of (2.37) includes only one term namely for the ith machine

$$B_{ij} E_i V_{i+n} \sin(\phi_i - \phi_{i+n}) \tag{2.38}$$

Load Flow Equations

1) The active power load is assumed to be the sum of voltage dependent component and a frequency dependent component, i.e.,

$$P_{\ell i} = P_i(V_i) + D_i \dot{\phi}_i, \qquad i = n+1, \dots n+N \tag{2.39}$$

where $\dot{\phi}_i$ represents the departure of the local frequency (as seen by load at i^{th} bus) from the nominal one.

2) The reactive power is assumed to be voltage dependent only

$$Q_{\ell i} = Q_i(V_i)\ , \qquad i = n+1, \ldots n+N \tag{2.40}$$

The load flow equations are now written as

$$D_i \dot{\phi}_i + P_i(V_i) + \sum_{j=1}^{n+N} B_{ij} V_i V_j \sin(\phi_i - \phi_j) = 0\ , \ i = n+1, \ldots n+N \tag{2.41}$$

$$Q_i(V_i) - \sum_{j=1}^{n+N} B_{ij} V_i V_j \cos(\phi_i - \phi_j) = 0\ , \ i = n+1, \ldots n+N \tag{2.42}$$

It may be noted that due to the sparse structure of the system the summation terms may contain only a few terms.

The above load modeling is quite general and (2.41) and (2.42) together with the swing equations (2.37), leads to a set of differential algebraic equations of the form

$$\dot{x} = f(x,y) \tag{2.43}$$

$$0 = g(x,y) \tag{2.44}$$

The various models appearing in the literature can be explained as follows.

1. Van-Cutsem and Ribbens-Pavella (1985)

In this formulation, real power loads are both voltage and frequency dependent and reactive power loads are only voltage dependent. If $D_i \neq 0$ for all i, then $x = (\delta^T, \omega^T, \psi^T)^T$ and $y = (V_{n+1}, \ldots V_{n+N})^T$. If some of the D_i's are zero, then

$$\begin{aligned} x &= (\delta^T, \omega^T, \text{some } \psi_i\text{'s})^T \\ y &= (\text{the other } \psi_i\text{'s}, V_{n+1}, \ldots V_{n+N})^T \end{aligned} \tag{2.45}$$

2) Padiyar and Sastry (1987)

Here the real and reactive power loads are only voltage dependent. Hence,

$$x = (\delta^T, \omega^T)^T$$
$$y = (\psi^T, V_{n+1}, \dots V_{n+N})^T \tag{2.46}$$

This is a special case of (1) with $D_i \equiv 0$ for all i.

3) Narasimhamurthy and Musavi (1984)

They assumed real power of the loads as constant while the reactive power is voltage dependent. Hence,

$$x = [\delta^T, \omega^T]^T \tag{2.47}$$
$$y = [\psi^T, V_{n+1}, \dots V_{n+N}]^T$$

4) Tsolas, Arapostathis and Varaiya (1985)

In this paper at the generator terminals, both the real and reactive power loads are assumed zero while at the other buses (i = 2n+1, ... n+N) both real and reactive powers are assumed constant. Hence,

$$x = [\delta^T, \omega^T]T$$
$$y = [\psi^T, V_{n+1}, \dots V_{n+N}]^T$$

5) Bergen and Hill Model (1981)

In this model all buses in the transmission network are considered as P-V nodes and Q at these nodes is not explicitly considered, i.e., there is no algebraic equation for reactive power at the buses. Hence, $|V_i|$ (i = n+1, ... n+N) is constant and we have real power loads as frequency dependent. Hence

$$x = [\delta^T, \omega^T, \psi^T]^T \tag{2.49}$$

There is no y vector and hence there are no algebraic equations.

2.5.2 Structure preserving energy functions

We now proceed to construct the energy function for the most general case

(1). The treatment follows on the lines of Van-Cutsem and Ribbens-Pavella (1985). Two cases are distinguished: (I) there is no frequency dependent load, (II) there is at least one frequency dependent load.

(I) No Frequency Dependent Loads

The system equations are

$$\dot{\delta}_i = (\omega_i - \omega_s) \tag{2.50}$$

$$M_i \dot{\omega}_i = P_{mi} - \sum_{j=1}^{n+N} B_{ij} V_i V_j \sin(\phi_i - \phi_j) \qquad i = 1, 2 \ldots n \tag{2.51}$$

$$P_i(V_i) = - \sum_{j=1}^{n+N} B_{ij} V_i V_j \sin(\phi_i - \phi_j) \qquad i = n+1, \ldots n+N \tag{2.52}$$

$$Q_i(V_i) = \sum_{j=1}^{n+N} B_{ij} V_i V_j \cos(\phi_i - \phi_j) \qquad i = n+1, \ldots n+N \tag{2.53}$$

State Equations in COI Reference Frame

We adopt the COI notation because it enables easier analytic calculations. The COI of a n-machine system is defined as

$$\delta_o = \frac{1}{M_T} \sum_{i=1}^{n} M_i \delta_i \text{ where } M_T = \sum_{i=1}^{n} M_i . \tag{2.54}$$

It follows that

$$\omega_o = \frac{1}{M_T} \sum_{i=1}^{n} M_i \omega_i , \quad \dot{\omega}_o = \frac{1}{M_T} \sum_{i=1}^{n} M_i \dot{\omega}_i \tag{2.55}$$

The machine rotor angles and bus phase angles referred to the COI are

$$\tilde{\phi}_i = \phi_i - \delta_o = \begin{cases} \tilde{\delta}_i = \delta_i - \delta_o & i = 1, 2, \ldots n \\ \tilde{\psi}_i = \psi_i - \delta_o & i = n+1, \ldots n+N \end{cases} \tag{2.56}$$

Hence

$$\dot{\tilde{\phi}}_i = \dot{\phi}_i - \dot{\delta}_o = \begin{cases} \tilde{\omega}_i = \omega_i - \omega_o & i = 1, 2, \ldots n \\ \dot{\tilde{\psi}}_i = \dot{\psi}_i - \omega_o & i = n+1, \ldots n+N \end{cases} \tag{2.57}$$

Summing up the n swing Eqs. (2.51)

$$\sum_{i=1}^{n} M_i \dot{\omega}_i = \sum_{i=1}^{n} P_{mi} - \sum_{i=1}^{n} \sum_{j=1}^{n+N} B_{ij} V_i V_j \sin(\phi_i - \phi_j) \tag{2.58}$$

Summing all the real power equations (2.52), we get

$$\sum_{i=n+1}^{n+N} P_i(V_i) = -\sum_{i=n+1}^{n+N} \sum_{j=1}^{n+N} B_{ij} V_i V_j \sin(\phi_i - \phi_j) \tag{2.59}$$

Adding (2.58) and (2.59) we get

$$\sum_{i=1}^{n} M_i \dot{\omega}_i = \sum_{i=1}^{n} P_{mi} - \sum_{i=n+1}^{n+N} P_i(V_i) - \sum_{i=1}^{n+N} \sum_{j=1}^{n+N} B_{ij} V_i V_j \sin(\phi_i - \phi_j) \tag{2.60}$$

The third term on the right-hand side of this equation is readily verified to be equal to zero since the network is lossless. Hence using Eq. (2.55)

$$M_T \dot{\omega}_o = \sum_{i=1}^{n} P_{mi} - \sum_{i=n+1}^{n+N} P_i(V_i) \triangleq P_{COI} \tag{2.61}$$

We can now rewrite (after some algebra) the swing equations (2.50) and (2.51) in the COI reference frame as

$$\begin{aligned} \dot{\tilde{\phi}}_i &= \tilde{\omega}_i \\ M_i \dot{\tilde{\omega}}_i &= P_{mi} - \sum_{j=1}^{n+N} B_{ij} V_i V_j \sin(\tilde{\phi}_i - \tilde{\phi}_j) - \frac{M_i}{M_T} P_{COI} \end{aligned} \tag{2.62}$$

The load flow equations (2.52) and (2.53) remain the same except that ϕ_i and ϕ_j get replaced by $\tilde{\phi}_i$ and $\tilde{\phi}_j$.

Setting up the first integral

(i) Dividing Eq. (2.53) by V_i gives

$$\frac{Q_i(V_i)}{V_i} = \sum_{j=1}^{n+N} B_{ij} V_j \cos(\tilde{\phi}_i - \tilde{\phi}_j)$$

$$= B_{ii} V_i + \sum_{\substack{j=1 \\ \neq i}}^{n+N} B_{ij} V_j \cos \tilde{\phi}_{ij} \tag{2.63}$$

where $\tilde{\phi}_{ij} = \tilde{\phi}_i - \tilde{\phi}_j$.

Multiply the expression (2.63) by $\dot{V}_i$ and sum over all the load buses. Since $\dot{V}_i \equiv 0$ at all internal buses, the summation involving B_{ij} terms can be done over all buses 1, ... n+N. This results in

$$\sum_{i=n+1}^{n+N} \frac{Q_i(V_i)}{V_i} \dot{V}_i = \sum_{i=n+1}^{n+N} B_{ii} V_i \dot{V}_i + \sum_{i=1}^{n+N} \sum_{\substack{j=1 \\ \neq i}}^{n+N} B_{ij} \dot{V}_i V_j \cos \tilde{\phi}_{ij}$$

$$= \sum_{i=n+1}^{n+N} B_{ii} V_i \dot{V}_i + \sum_{i=1}^{n+N-1} \sum_{j=i+1}^{n+N} B_{ij} (\dot{V}_i V_j + V_i \dot{V}_j) \cos \tilde{\phi}_{ij} \tag{2.64}$$

since $B_{ij} = B_{ji}$ $(j \neq i)$ and $\cos \tilde{\phi}_{ij} = \cos \tilde{\phi}_{ji}$.

(ii) Multiply the swing equations (2.62) by $\tilde{\omega}_i = \dot{\tilde{\phi}}_i$ and sum over the n machine equations. Noting that $\sum_{i=1}^{n} M_i \tilde{\omega}_i = 0$, we get

$$\sum_{i=1}^{n} M_i \dot{\tilde{\omega}}_i \tilde{\omega}_i = \sum_{i=1}^{n} P_{mi} \dot{\tilde{\phi}}_i - \sum_{i=1}^{n} \sum_{j=1}^{n+N} B_{ij} V_i V_j \sin \tilde{\phi}_{ij} \dot{\tilde{\phi}}_i \tag{2.65}$$

$$i = 1, 2, \ldots n$$

Similarly we multiply Eq. (2.52) by $\dot{\tilde{\phi}}_i$ and sum over the N active power load equations

$$\sum_{i=n+1}^{n+N} P_i(V_i) \dot{\tilde{\phi}}_i = - \sum_{i=n+1}^{n+N} \sum_{j=1}^{n+N} B_{ij} V_i V_j \sin \tilde{\phi}_{ij} \dot{\tilde{\phi}}_i \tag{2.66}$$

$$i = n+1, \ldots, n+N$$

Adding Eq. (2.65) and (2.66) we get

$$\sum_{i=1}^{n} M_i \tilde{\omega}_i \dot{\tilde{\omega}}_i + \sum_{i=n+1}^{n+N} P_i (V_i) \dot{\tilde{\phi}}_i = \sum_{i=1}^{n} P_{mi} \dot{\tilde{\phi}}_i - \sum_{i=1}^{n+N-1} \sum_{j=i+1}^{n+N} B_{ij} V_i V_j \sin \tilde{\phi}_{ij} \dot{\tilde{\phi}}_{ij} \tag{2.67}$$

A careful examination of the second term on the right-hand side of (2.65) and the right-hand side of (2.66) will verify the result in (2.67).

Adding this to the reactive component Eq. (2.64) one can easily verify

$$\frac{d}{dt}\left[\frac{1}{2}\sum_{i=1}^{n} M_i \tilde{\omega}_i^2 - \sum_{i=1}^{n} P_{mi} \tilde{\phi}_i + \sum_{i=n+1}^{n+N} \int \frac{Q_i(V_i)}{V_i} d V_i - \frac{1}{2}\sum_{i=n+1}^{n+N} B_{ii} V_i^2 - \sum_{i=1}^{n+N-1} \sum_{j=i+1}^{n+N} B_{ij} V_i V_j \cos \tilde{\phi}_{ij}\right] + \sum_{i=n+1}^{n+N} P_i(V_i) \dot{\tilde{\phi}}_i = 0 \tag{2.68}$$

The fifth term in the square brackets, when differentiated using the chain rule, gives rise to the various terms of (2.64) and (2.67). The term involving $P_i(V_i)$ causes some problems. We consider two cases.

(a) $P_i(V_i) = P_i = \text{constant}$

In this case the first integral from (2.68) is

$$V(\tilde{\omega}, \tilde{\phi}, V) = V_k(\tilde{\omega}) + V_{p1} (\tilde{\phi},V) + V_{p2} (\tilde{\phi}) + K \tag{2.69}$$

where $V_k(\tilde{\omega}) = \frac{1}{2} \sum_{i=1}^{n} M_i \tilde{\omega}_i^2$ and K is an arbitrary constant chosen such that at the post-fault stable equilibrium point ($\tilde{\omega}=0$, $\tilde{\phi}=\tilde{\phi}^s$, $V=V^s$), $V(\tilde{\omega},\tilde{\phi},V) \equiv 0$. Hence $K = -V_{p1} (\tilde{\phi}^s, V^s) - V_{p2} (\tilde{\phi}^s)$. Therefore

$$V_{p1} (\tilde{\phi},V) = -\sum_{i=1}^{n} P_{mi}(\tilde{\phi}_i - \tilde{\phi}_i^s) + \sum_{i=n+1}^{n+N} \int_{V_i^s}^{V_i} \frac{Q_i(V_i)}{V_i} dV_i - \frac{1}{2} \sum_{i=n+1}^{n+N} B_{ii} \left(V_i^2 - (V_i^s)^2\right)$$

$$- \sum_{i=1}^{n+N-1} \sum_{j=i+1}^{n+N} B_{ij} \left(V_i V_j \cos \tilde{\phi}_{ij} - V_i^s V_j^s \cos \tilde{\phi}_{ij}^s\right) \tag{2.70}$$

and

$$V_{p2}(\tilde{\phi}) = \sum_{i=n+1}^{n+N} P_i (\tilde{\phi}_i - \tilde{\phi}_i^s). \tag{2.71}$$

At a stable equilibrium point all $\omega_i \to \omega_o$ so that $\tilde{\omega}_i^s = 0$. V can be interpreted as the global energy of the system and V_k and (V_{p1}, V_{p2}) are, respectively, the transient kinetic energy and potential energy components. An equivalent expression for V_{p1} is obtained by combining the third and fourth terms in (2.70) as

$$V_{p1}(\tilde{\phi},V) = - \sum_{i=1}^{n} P_{mi} (\tilde{\phi}_i - \tilde{\phi}_i^s) + \sum_{i=n+1}^{n+N} \int_{V_i^s}^{V_i} \frac{Q_i(V_i)}{V_i} dV_i$$

$$- \frac{1}{2} \sum_{i=1}^{n+N} \sum_{j=1}^{n+N} B_{ij} \left[V_i V_j \cos \tilde{\phi}_{ij} - V_i^s V_j^s \cos \tilde{\phi}_{ij}^s\right] \tag{2.72}$$

Using (2.53), it is easy to show that the last term is half the sum of the reactive powers injected into the extended system and consequently it is half the sum of the reactive power dissipated in its branches.

(b) $P_i(V_i)$ is voltage dependent

It is not possible to get a first integral in this case. We may however approximate it in several different ways.

1) Treat it as a path dependent integral

$$\sum_{i=n+1}^{n+N} \int_o^t P_i(V_i) \dot{\tilde{\phi}}_i \, dt = \sum_{i=n+1}^{n+N} \int_{\tilde{\phi}_i^s}^{\tilde{\phi}_i} P_i(V_i) \, d\tilde{\phi}_i \tag{2.73}$$

2) Replace $P_i(V_i) \dot{\tilde{\phi}}_i$ by $\frac{d}{dt} \left[P_i(V_i) \tilde{\phi}_i\right] - \frac{dP_i(V_i)}{dV_i} \dot{V}_i \tilde{\phi}_i$ and neglect the second term. Thus,

$$V_{app}(\tilde{\omega},\tilde{\phi},V) = V_k(\tilde{\omega}) + V_{p1}(\tilde{\phi}, V) + \sum_{i=n+1}^{n+N} \left[P_i(V_i)\,\tilde{\phi}_i - P_i(V_i^s)\tilde{\phi}_i^s\right] \tag{2.74}$$

It is clear that

$$\frac{d}{dt} V_{app} = \sum_{i=n+1}^{n+N} \frac{dP_i(V_i)}{dV_i} \dot{V}_i\, \tilde{\phi}_i \tag{2.75}$$

and this may be small in magnitude and < 0.

(II) Frequency Dependent Loads

The presence of frequency dependent loads produces some difficulty in obtaining a first integral in terms of $\tilde{\omega}_i$ for kinetic energy. Hence, we take the model as the swing equation in ω_i variables.

$$\dot{\phi}_i = (\omega_i - \omega_s)$$

$$M_i\, \dot{\omega}_i = P_{mi} - \sum_{j=1}^{n+N} B_{ij}\, V_i\, V_j \sin \phi_{ij} \qquad i = 1, 2 \ldots n \tag{2.76}$$

The equations at the transmission network buses are

$$D_i\, \dot{\phi}_i + P_i(V_i) = -\sum_{j=1}^{n+N} B_{ij}\, V_i\, V_j \sin \phi_{ij} \tag{2.77}$$

$$i = n+1,\ldots,n+N$$

$$Q_i(V_i) = \sum_{j=1}^{n+N} B_{ij}\, V_i\, V_j \cos \phi_{ij} \tag{2.78}$$

Equations (2.77) and (2.78) can be written as

$$P_i(V_i) + D_i\, \dot{\tilde{\phi}}_i = -\sum_{j=1}^{n+N} B_{ij}\, V_i\, V_j \sin \tilde{\phi}_{ij} - D_i\, \omega_o \qquad i = n+1, \ldots n+N \tag{2.79a}$$

$$Q_i(V_i) = \sum_{j=1}^{n+N} B_{ij}\, V_i\, V_j \cos \tilde{\phi}_{ij} \tag{2.79b}$$

In (2.76), ϕ_{ij} can be replaced by $\tilde{\phi}_{ij}$.

The motion of the COI is obtained by summing all the swing equations (2.76) and the real power balance equation (2.79a). Thus,

$$M_T \dot{\omega}_o + D_T \omega_o = - \sum_{i=n+1}^{n+N} D_i \dot{\tilde{\phi}}_i + \left[\sum_{i=1}^{n} P_{mi} - \sum_{i=n+1}^{n+N} P_i(V_i) \right] \tag{2.80}$$

with $D_T = \sum_{i=n+1}^{n+N} D_i$. The procedure is the same as in the previous case except that Eq. (2.76) is multiplied by ω_i (instead of $\tilde{\omega}_i$). After some analytical calculations (details omitted) we get

$$\begin{aligned} \frac{d}{dt} \Bigg[& \sum_{i=1}^{n} M_i \omega_i^2 - \sum_{i=1}^{n} P_{mi} \tilde{\phi}_i + \sum_{i=n+1}^{n+N} \int \frac{Q_i(V_i)}{V_i} dV_i \\ & - \sum_{i=1}^{n+N-1} \sum_{j=i+1}^{n+N} B_{ij} V_i V_j \cos \tilde{\phi}_{ij} - \frac{1}{2} \sum_{i=n+1}^{n+N} B_{ii} V_i^2 \Bigg] \\ & + \sum_{i=n+1}^{n+N} P_i(V_i) \dot{\tilde{\phi}}_i \\ & = - \sum_{i=n+1}^{n+N} D_i \dot{\phi}_i^2 + \omega_o \left[\sum_{i=1}^{n} P_{mi} - \sum_{i=n+1}^{n+N} P_i(V_i) \right] \end{aligned} \tag{2.81}$$

Again, two cases are to be considered.

(i) Active power loads are constant, i.e., $P_i(V_i) = P_i$

Suppose there exists an equilibrium such that $\dot{\omega}_i^s = 0$ and $\dot{\tilde{\phi}}_i^s = 0$. This requires from (2.80) that

$$\sum_{i=1}^{n} P_{mi} - \sum_{i=n+1}^{n+N} P_i = 0 \tag{2.82}$$

With these assumptions we have the same energy function V as before (Eqs. (2.69)-(2.71)) except that ω_i replaces $\tilde{\omega}_i$ such that $\dot{V}(\omega,\tilde{\phi},V) = -\sum_{i=n+1}^{n+N} D_i \dot{\phi}_i^2$. This energy function generalizes the energy function of Bergen and Hill (1981).

(ii) Active power loads are voltage dependent.

The same approximation as in the previous case has to be made regarding the term $P_i(V_i)\dot{\tilde{\phi}}_i$, i.e.,

$$P_i(V_i)\dot{\tilde{\phi}}_i = \frac{d}{dt}\left[P_i(V_i)\tilde{\phi}_i\right] - \frac{dP_i(V_i)}{dV_i}\dot{V}_i\tilde{\phi}_i \tag{2.83}$$

Assuming that an equilibrium exists, i.e.,

$$\sum_{i=n+1}^{n+N} P_{mi} - \sum_{i=n+1}^{n+N} P_i(V_i^s) = 0 \tag{2.84}$$

The new energy function denoted by V_{app} is the same as before except that $P_i(V_i)$ replaces P_i and

$$\frac{d}{dt}V_{app} = -\sum_{i=n+1}^{n+N} D_i\dot{\phi}_i^2 + \sum_{i=n+1}^{n+N}\frac{dP_i(V_i)}{dV_i}\dot{V}_i\tilde{\phi}_i \tag{2.85}$$

If the right-hand side in Eq. (2.85) is ≤ 0 the equilibrium is stable.

2.6 Alternative Form of the Structure Preserving Energy Function

While the energy functions discussed so far are concise, an alternative form which separates the terms due to the transient reactances of generators in the expression for $V_{P1}(\tilde{\phi},V)$ is also possible. In Sastry (1984) the energy function is developed in a different manner but equivalent to the one in the previous section. The expression for $V_{p1}(\tilde{\phi},V)$ is given by

$$V_{P1}(\tilde{\phi},V) = -\sum_{i=1}^{n} P_{mi}(\tilde{\phi}_i - \tilde{\phi}_i^s) + \sum_{i=n+1}^{n+N}\int_{V_i^s}^{V_i}\frac{Q_i(V_i)}{V_i}dV_i$$

$$- \frac{1}{2} \sum_{i=n+1}^{n+N} \sum_{j=n+1}^{n+N} B_{ij} (V_i V_j \cos \tilde{\phi}_{ij} - V_i^s V_j^s \cos \tilde{\phi}_{ij}^s)$$

$$+ \sum_{i=1}^{n} \Bigg[\{(V_i^2 + V_{i+n}^2 - 2 V_i V_{i+n} \cos (\tilde{\phi}_i - \tilde{\phi}_{i+n}))$$

$$- (V_i^{s^2} + V_{i+n}^{s^2} - 2 V_i^s V_{i+n}^s \cos (\tilde{\phi}_i^s - \tilde{\phi}_{i+n}^s))\} \Bigg] \frac{1}{2x'_{di}} \qquad (2.86)$$

Recognizing the fact that in this formulation the Y_{BUS} pertains to the network alone, the equivalence of (2.86) to (2.72) can be established. The way of handling real power loads remains the same as in the previous section.

2.7 Positive Definiteness of the Energy Integral

In order to justify the existence of a region of attraction around the post-fault stable equilibrium point, it is necessary to show that $V(x) > 0$ and $\dot{V}(x) \leq 0$ and that the only invariant set inside the region is the origin. In power system stability literature with internal node representation and transfer conductances included, it is not possible to prove $V(x) > 0$ mathematically. However, from a physical point of view, energy is injected into the system for $t > 0$ until $t = t_{c\ell}$ so that $V(x) > 0$. Hence, based on simulations and experience, the sign definite properties of $V(x)$ and $\dot{V}(x)$ have been shown to be true. However, in the case of structure preserving model with nonlinear loads use of Moore - Anderson theorem in conjunction with Lure´ representation of the system equations, it is possible to show that the energy integral is indeed a Lyapunov function. This was first shown by Bergen and Hill (1981) and extended by Hill and Chong (1985) to detailed load models. Since it is necessary to have a minimal state space representation to apply the Lyapunov theory, the internodal angles and speeds of machines are chosen as state variables. In COI representation, the system of equations become over determined. However, Chong (1986) shows that the energy integral in terms of internodal angles is equivalent to the energy integral in terms of the COI notation.

In the work of Tsolas et al. (1985), they consider a structure preserving model with zero loads at generator buses and constant P-Q loads at the other network buses. They construct the first integral as the energy function. Except for change of variables $\nu = \log V$ for the transmission network voltages, the results appear to be same as before. They separate the potential energy terms due to generator reactances and the transmission network as in Section 2.6. They consider both the classical and the flux decay model. This paper is significant in two respects: (i) in the matter of handling loads in a structure preserving framework and (ii) in terms of characterization of the region of stability. While (ii) will be discussed in a later chapter, we will explain (i) for the classical model.

2.8 Tsolas-Araposthasis-Varaiya Model (Classical Machine Representation)(1985)

In this model, instead of choosing a COI as reference, one of the buses whose frequency is constant ω_s and the bus voltage is constant (=1 p.u.) with phase angle zero is considered as reference (infinite bus). Without loss of generality, let there be n+N+1 including the infinite bus. The (2n+1)th bus is the infinite bus and the physical buses are ordered n+1, ... n+N as before. At the terminal buses of the generators (n+1) ... 2n+1, there are no loads whereas at the buses 2n+2, ... n+N+n the loads are constant P and constant Q. All phase angles are measured with respect to bus (2n+1), whose phase angle is zero.

The differential-algebraic equations are

$$\dot{\delta} = \omega - \omega_s \tag{2.87}$$

$$M\dot{\omega} = P_m - D(\omega-\omega_s) - f(\delta,\phi,V) \tag{2.88}$$

$$0 = P^d - g(\delta,V,\phi) \tag{2.89}$$

$$0 = Q^d - h(\delta,V,\phi) \tag{2.90}$$

where (2.89) and (2.90) are the load flow equations at the buses n+1, ... n+N. δ, ω, P_m, f, ϕ, V, P^d, g, Q^d and h are vectors of appropriate order and M, D are diagonal matrices. P, Q are the injected powers.

$f_i\ (\delta,\phi,V)$ is given by $f_i = \frac{E_i V_i}{X'_{di}} \sin(\delta_i - \phi_i)$ where E_i, δ_i refer to the internal bus and V_i, ϕ_i refer to the corresponding physical bus. The load flow equations are

$$0 = g_i\ (\delta,E,V,\phi)$$

$$= E_i \frac{V_i}{X'_{di}} \sin(\phi_i - \delta_i) + \sum_{\substack{j=n+1 \\ \neq i}}^{2n+1} B_{ij} V_i V_j \sin(\phi_i - \phi_j)$$

$$+ \sum_{k=2n+2}^{n+N+1} B_{ik} V_i V_k \sin(\phi_i - \phi_k) \tag{2.91}$$

$$i = n+1, \ldots\ldots 2n$$

$$0 = h_i\ (\delta,E,V,\phi)$$

$$= \frac{V_i^2}{X'_{di}} - E_i \frac{V_i \cos(\phi_i - \delta_i)}{X'_{di}} - B_{ii} V_i^2 - \sum_{\substack{j=n+1 \\ \neq i}}^{2n+1} B_{ij} V_i V_j \cos(\phi_i - \phi_j)$$

$$- \sum_{k=2n+2}^{n+N+1} B_{ik} V_i V_k \cos(\phi_i - \phi_k),\quad i = n+1, \ldots 2n \tag{2.92}$$

E_i in (2.91) and (2.92) pertain to the internal node voltage corresponding to terminal bus i. At the kth load bus

$$P_k^d = g_k(\delta,\phi,V)$$

$$= \sum_{i=n+1}^{2n+1} B_{ki} V_k V_i \sin(\phi_k - \phi_i)$$

$$+ \sum_{\ell=2n+2}^{n+N+1} B_{k\ell} V_k V_\ell \sin(\phi_k - \phi_\ell) \quad , \quad k = 2n+2, \ldots n+N+1 \tag{2.93}$$

$$Q_k^d = h_k(\delta,\phi,V) = - B_{kk} V_k^2 - \sum_{i=n+1}^{2n+1} B_{ki} V_k V_i \sin(\phi_k - \phi_i)$$

$$- \sum_{\ell=2n+2}^{n+N} B_{k\ell} V_k V_\ell \cos(\phi_k - \phi_\ell) \quad , \quad k = 2n+2, \ldots\ldots n+N+1 \tag{2.94}$$

Define $\nu_i = \log V_i$, $i = n+1, \ldots n+N+1$. The energy function is

$$V(\delta,\omega,\phi,\nu)$$

$$= \frac{1}{2} \sum_{i=1}^{n} M_i(\omega_i - \omega_s)^2 + V_2(\delta,\phi,\nu) + K \tag{2.95}$$

where K = constant

and $V_2(\delta,\phi,\nu)$

$$= - \sum_{i=1}^{n} P_{mi}\,\delta_i - \sum_{i=2n+2}^{n+N+1} P_i^d \phi_i - \sum_{i=2n+2}^{n+N+1} Q_i^d \nu_i$$

$$- \sum_{i=1n+1}^{2n} \frac{E_i e^{\nu i} \cos(\delta_i - \phi_i)}{X'_{di}} - \sum_{i=n+1}^{2n} (B_{ii} - \frac{1}{X'_{di}}) \frac{e^{2\nu i}}{2}$$

$$- \sum_{i<j}^{2n+1} B_{ij}\, e^{\nu i + \nu j} \cos(\phi_i - \phi_j) \tag{2.96}$$

$$- \sum_{i=n+1}^{2n+1} \sum_{k=2n+2}^{n+N+1} B_{ik}\, e^{\nu i + \nu k} \cos(\phi_i - \phi_k)$$

$$- \sum_{k<\ell}^{n+N+1} B_{k\ell}\, e^{\nu k + \nu \ell} \cos(\phi_k - \phi_\ell) - \sum_{k=2n+2}^{n+N+1} B_{kk} \frac{e^{2\nu k}}{2}$$

In (2.96) E_i, δ_i refer to the internal nodes connected to physical bus i. It is the practice to make V = 0 at the stable equilibrium point. This is done by subtracting $V_2\ (\delta^s,\phi^s,V^s)$ from (2.95). It can be verified that

$$-\frac{\partial V_2}{\partial \delta_i} = P_{mi} - f_i(\delta,\phi,\nu) \qquad i=1,\ 2\ \dots\ n \tag{2.97}$$

$$-\frac{\partial V_2}{\partial \phi_i} = -\ g_i(\delta,\phi,\nu) \tag{2.98}$$

$$-\frac{\partial V_2}{\partial \nu_i} = -\ h_i(\delta,\phi,\nu)$$

$$-\frac{\partial V_2}{\partial \phi_k} = P_k^{\ d} - g_k(\delta,\nu,\phi) \tag{2.100}$$

$$k=2n+1,\ \dots\ n+N$$

$$-\frac{\partial V_2}{\partial \nu_k} = Q_k^{\ d} - h_k(\delta,\nu,\phi) \tag{2.101}$$

The constant K in (2.95) can be adjusted to make $V(\delta^s,\omega_s,\phi^s,\nu^s) = 0$. Using these it can be shown that

$$\frac{dV}{dt} = -\sum_{i=1}^{n} D_i(\omega_i-\omega_s)^2 \leq 0. \tag{2.102}$$

A generalization along these lines for detailed models is contained in the work by Padiyar and Sastry (1987) and Padiyar and Ghosh (1989).

2.9 Conclusion

In this chapter we have considered energy functions for internal node and structure preserving models and have shown that except when the real power load is voltage dependent a first integral can be constructed. It has been shown that it is indeed a Lyapunov function.

Chapter 3

REDUCED ORDER ENERGY FUNCTIONS

3.1 Introduction

In the previous chapter we discussed the system-wide energy functions using the classical model for both the internal node model and the structure preserving model. It is a well known fact, observed through numerous simulations that depending on the fault it is always a single machine or a group of machines that pull away from the rest of the machines. In some instances there may be several groups of machines pulling out from the rest of the system. A group of machines may even decelerate. These cases are illustrated in Figs. 3.1a, b, and c.

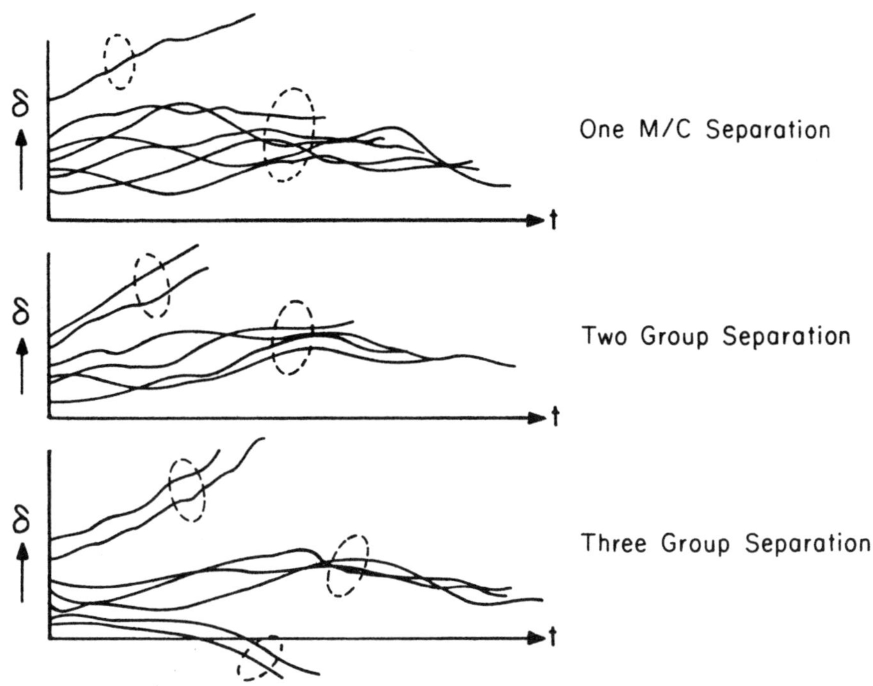

Figure 3.1 Different modes of instability

These cases occur depending on the nature and location of the fault and how tightly or weakly the machines are coupled. In these cases, if we have an a-priori information as to the mode of instability or can infer it on-line, a reduced order energy function would suffice. In the case of Fig. 3.1a, an individual machine energy function (IMEF) was proposed by Michel, Fouad and Vittal (1983). This is a generalization of a single machine infinite bus concept. In the case of Fig. 3.1b, the machines pulling away can be combined into one equivalent machine and the rest of the machines can be treated as in the case of the individual energy function method. Alternatively, the rest of the machines can be treated as a single equivalent machine and a two machine energy function can be proposed. Finally, in the case of Fig. 3.1c, three groups of machines can be combined into three equivalent machines and hence a three machine energy function analysis is possible. This concept can be extended to any number of the groups of machines. The cutset energy function (CSEF) proposed by Chandrasekhar and Hill (1986) and the works of Xue et al. (1988) and Lemmon et al. (1988) deal with cases such as in Fig. 3.1b, while the time scale energy function proposed by Pai et al. (1986) deals with a general class of instability such as in Fig. 3.1c. The concept of time scale energy function is based on the notion that a power system inherently consists of strong and weak connections. This leads to the concept of slow-coherency grouping where the machines which are grouped together are strongly interconnected with each other. In this chapter we discuss these various reduced order energy functions.

3.2 Individual Machine and Group Energy Function

3.2.1 Individual machine energy function (Michel et al. (1983))

The expression for individual machine energy function for the internal node model has been derived in Chapter 2 as

$$V_i = \frac{1}{2} M_i \tilde{\omega}_i^2 - P_i(\theta_i - \theta_i^s) + \sum_{\substack{j=1 \\ \neq i}}^{n} C_{ij} \int_{\theta_i^s}^{\theta_i} \sin \theta_{ij} \, d\theta_i$$

$$+ \sum_{\substack{j=1 \\ \neq i}}^{n} D_{ij} \int_{\theta_i^s}^{\theta_i} \cos \theta_{ij} \, d\theta_i + \frac{M_i}{M_T} \int_{\theta_i^s}^{\theta_i} P_{COI} \, d\theta_i \tag{3.1}$$

$$= V_{K \cdot E} (\tilde{\omega}_i) + V_{P \cdot E} (\theta)$$

If the mode of instability is such that a particular machine (critical machine) is likely to pull away from the rest of the system, then the V_i for that particular machine can be used as the energy function instead of the system wide energy function. However, the faulted equations for all the machines need to be integrated to know the θ_i's of machines other than the critical machine. But it is more accurate than the infinite bus equivalent. The individual machine energy function, as expected, should work well for faults near the generator. Generally, that generator becomes unstable if the fault is not cleared in time. Sometimes a close by generator may become unstable first. Analogous to the single machine case, $V_{PEmax,i}$ along the faulted trajectory can be taken as $V_{cr,i}$. For each machine this quantity can be computed and stored. The critical machine is the one in which $V_i = V_{cr,i}$ first along the fault-on trajectory. For faults away from the generator terminal this method may not give the correct mode of instability.

3.2.2 Group energy function

If more than one machine pulls away from the rest of the system, then we need to compute the group energy function. A physical assumption is made that there is no transfer of energy between machines in the critical group. This fact can be interpreted as subtracting the P•E due to intermachine oscillations in the group from the sum of individual P•E's of the critical machines. We can obtain V_{group} easily as follows. Suppose machines 1 and 2 form the critical group. Then the potential energy of the group with respect to the rest of the system is given by

$$V_{P.E1,2} = \left(V_{P.E1} + V_{P.E2}\right)\Bigg|_{\substack{C_{12} = C_{21} = 0 \\ D_{12} = D_{21} = 0}} \tag{3.2}$$

This has been explained alternatively by Vittal (1982) as follows:

Let the power flows from node i to j and j to i be given respectively as

$$P_{ij} = C_{ij} \sin \theta_{ij} + D_{ij} \cos \theta_{ij} \tag{3.3}$$

$$P_{ji} = C_{ji} \sin \theta_{ji} + D_{ji} \cos \theta_{ji} \tag{3.4}$$

The change in power flow ΔPE in branch ij is given by

$$\Delta PE = \int_{\theta_i^s}^{\theta_i} P_{ij}\, d\theta_i + \int_{\theta_j^s}^{\theta_j} P_{ji}\, d\theta_j \tag{3.5}$$

The potential energy of the group with respect to the rest of the system is given by

$$V_{P.E1,2} = V_{P.E1} + V_{P.E2} - (\Delta PE \text{ due to power flow between machines 1 and 2})$$

$$= - \sum_{i=1}^{2} P_i(\theta_i - \theta_i^s) + \sum_{i=1}^{2} \sum_{j=3}^{n} C_{ij} \int_{\theta_i^s}^{\theta_i} \sin \theta_{ij}\, d\theta_i$$

$$+ \sum_{i=1}^{2} \sum_{j=3}^{n} D_{ij} \int_{\theta_i^s}^{\theta_i} \cos \theta_{ij}\, d\theta_i + \frac{1}{M_T} \sum_{i=1}^{2} M_i \int_{\theta_i^s}^{\theta_i} P_{COI}\, d\theta_i \tag{3.6}$$

Note that this physical reasoning is equivalent to assuming that there is no exchange of P•E within a group. In general if machines 1, 2 ... k form the critical group and machines k+1, ...n belong to the intact group, then

$$V_{P.E1,2...k} = -\sum_{i=1}^{k} P_i(\theta_i-\theta_i^s) + \sum_{i=1}^{k}\sum_{j=k+1}^{n}\left[C_{ij}\int_{\theta_i^s}^{\theta i}\sin\theta_{ij}\,d\theta_i + D_{ij}\int_{\theta_i^s}^{\theta i}\cos\theta_{ij}\,d\theta_i\right] + \frac{1}{M_T}\sum_{i=1}^{k} M_i\int_{\theta_i^s}^{\theta i} P_{COI}\,d\theta_i \triangleq V_{P\bullet E(group)} \tag{3.7}$$

Now we compute the transient kinetic energy of the group with respect to the rest of the system. Let $M_{cr} = \sum_{i=1}^{k} M_i$ and $M_{sys} = \sum_{i=k+1}^{n} M_i$. Let $\tilde{\omega}_{cr}$ and $\tilde{\omega}_{sys}$ be the center of speeds of the critical and the intact group, respectively. Then the kinetic energy for the equivalent two-machine group is given as:

$$V_{K.E(eq)} = \frac{1}{2}\frac{M_{cr}\,M_{sys}}{M_{cr}+M_{sys}}(\tilde{\omega}_{cr}-\tilde{\omega}_{sys})^2 \tag{3.8}$$

Proof $\quad \tilde{\omega}_0 = (M_{cr}\,\tilde{\omega}_{cr} + M_{sys}\,\tilde{\omega}_{sys})/(M_{cr}+M_{sys})$

The sum of the transient kinetic energies of COI's of the critical group and the rest of the system is given by

$$V_{K.E(eq)} = \frac{1}{2}M_{cr}(\tilde{\omega}_{cr}-\tilde{\omega}_0)^2 + \frac{1}{2}M_{sys}(\tilde{\omega}_{sys}-\tilde{\omega}_0)^2 \tag{3.9}$$

Using the expression for $\tilde{\omega}_0$, it can be shown (algebra omitted) that

$$V_{K.E(eq)} = \frac{1}{2}\frac{M_{cr}\,M_{sys}}{M_{cr}+M_{sys}}(\tilde{\omega}_{cr}-\tilde{\omega}_{sys})^2 \tag{3.10}$$

The total transient energy of the group with respect to the rest of the system is given by

$$V_{group} = V_{K\bullet E(eq)} + V_{P\bullet E(group)} \tag{3.11}$$

where $V_{P\bullet E(group)}$ is given by (3.7). In computing (3.11), the connections between all the machines in the group and the rest of the system are involved. In the special case where the critical machine is i, the group V-function is given by (3.1) where $\frac{1}{2} M_i \tilde{\omega}_i^2$ is replaced by $\frac{1}{2} \frac{M_i M_{sys}}{(M_i + M_{sys})} (\tilde{\omega}_i - \tilde{\omega}_{sys})^2$.

The difficult question is that of answering a-priori which are the critical machines for a given fault. Once it is known, a two-machine equivalent can be found by considering the COI's of the two areas, i.e., the critical group and the rest of the system [Xue et al. (1988)]. The two-machine equivalent can then be reduced to a single machine infinite bus and has been explained in Kimbark (1948) and in the previous chapter. An alternative procedure which uses the system s.e.p. and controlling u.e.p. is given in Fouad and Vittal (1983).

3.3 Simplified Form of the Individual Machine Energy Function [Siddiqee and Peschon (1969)]

This technique can be considered as a simplification of the individual machine energy function or the two-machine equivalent by converting it to a single-machine infinite bus case for faults involving sudden loss of a heavily loaded line near a generator. These kinds of studies are done in early planning stages. With an internal node representation the swing equations are

$$M_i \frac{d^2\delta_i}{dt^2} = P_{mi} - \mathrm{Re}\left[|E_i| \sum_{j=1}^{n} (G_{ij} - jB_{ij}) |E_j| \angle \delta_i - \delta_j \right]$$

$$= P_{mi} - \mathrm{Re}\left[|E_i| |E_j| |Y_{ij}| e^{j\angle -90^\circ} e^{j\angle \delta_{ij}} \right] \quad \text{after neglecting } G_{ij} \text{ terms } i \neq j$$

$$= P_{mi} - |E_i^2| G_{ii} - \sum_{\substack{j=1 \\ \neq i}}^{n} |E_i| |E_j| |Y_{ij}| \cos(90^\circ - \delta_{ij})$$

$$= P_i - \sum_{\substack{j=1 \\ \neq i}}^{n} |E_i||E_j| B_{ij} \sin(\delta_{ij})$$

$$= P_i - \sum_{\substack{j=1 \\ \neq i}}^{n} C_{ij} \sin \delta_{ij} \tag{3.12}$$

where $P_i = P_{mi} - |E_i|^2 G_{ii}$

$C_{ij} = |E_i||E_j| B_{ij}$

Assumptions

1) Assume for simplicity $|E_i| = |E_j| = |E|$, $(i \neq j)$

$$M_i \frac{d^2\delta_i}{dt^2} = P_i - \sum_{\substack{j=1 \\ \neq i}}^{n} |E|^2 B_{ij} \sin \delta_{ij} \tag{3.13}$$

2) Assume that $\delta_j (j \neq i)$ are at their post-fault values δ_j^s. Then

$$M_i \frac{d^2\delta_i}{dt^2} = P_i - \sum_{\substack{j=1 \\ \neq i}}^{n} |E|^2 |B_{ij}| \sin(\delta_i - \delta_j^s) \tag{3.14}$$

It will be shown that the right-hand side of (3.14) can be expressed as

$$P_i - |y_i| \sin(\delta_i - \psi^s) = P_i - |y_i| \sin \theta_i \tag{3.15}$$

where $\theta_i = \delta_i - \psi^s$ is the angle between E_i and the vector sum of $\sum_{\substack{j=1 \\ \neq i}}^{n} |E|^2 B_{ij} \sin(\delta_i - \delta_j^s)$. ψ^s is the angle of this vector sum with respect to the synchronously rotating reference frame.

Proof In Fig. 3.2, the internal voltages $E_1, E_2 \ldots E_n$ are shown schematically and their phase angles are measured relative to the synchronous reference (not shown in the figure). If along each of these phasors we draw the vectors A_1, $A_2 \ldots$ such that $|A_j| = |E|^2 B_{ij} (j=1, 2 \ldots n, \neq i)$ then $\sum_{\substack{j=1 \\ \neq i}}^{n} |E|^2 B_{ij} \sin(\delta_i - \delta_j^s)$

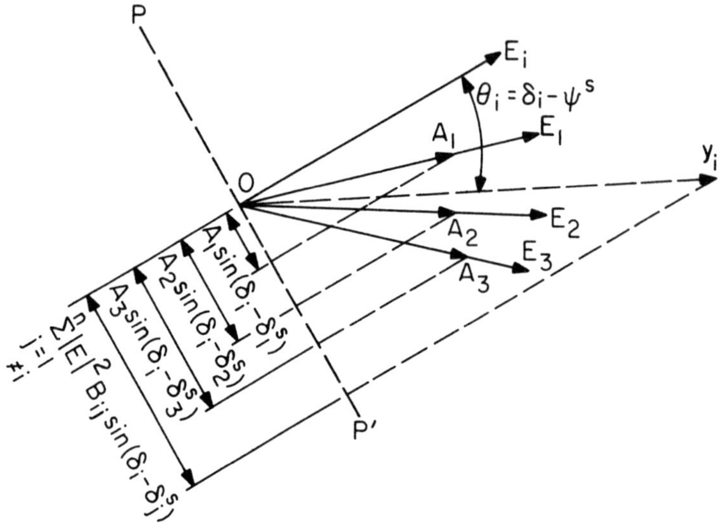

Fig. 3.2 Phasor representation $\sum_j |E|^2 B_{ij} \sin(\delta_i - \delta_j^s)$

is the sum of the projection of A_j's on a line POP' perpendicular to E_i. The phasor sum of $A_j (j \neq i)$ gives a phasor y_i and the phase angle E_i with respect to y_i is $\theta_i = \delta_i - \psi^s$. The magnitude of y_i can be shown to be equal to

$$|y_i| = \sqrt{\sum_{\substack{j=1 \\ \neq i}}^{n} (|E|^2 B_{ij})^2 + 2 \sum_{\substack{j,k \\ j,k \neq i}} |E|^2 B_{ij} \, |E|^2 B_{ik} \cos(\delta_j^s - \delta_k^s)}$$

Note that the magnitude y_i is a function of B_{ij} and the relative angle differences of voltages not involving i. We thus have an equivalent single machine infinite bus case for machine i as

$$M_i \frac{d^2\theta_i}{dt^2} = P_i - |y_i| \sin \theta_i \tag{3.16}$$

since $$\frac{d^2\theta_i}{dt^2} = \frac{d^2\delta_i}{dt^2}$$

Equal area criterion or the energy function method can be applied to this system. In turn each machine can be converted to a similar equivalent. Sudden loss of a heavily loaded line from the machine bus gives rise to only a prefault and post-fault power angle curves with different values for y_i. Let these values of y_i be y_i^o and y_i^*, respectively. The power angle curves for machine i are shown in Fig. 3.3.

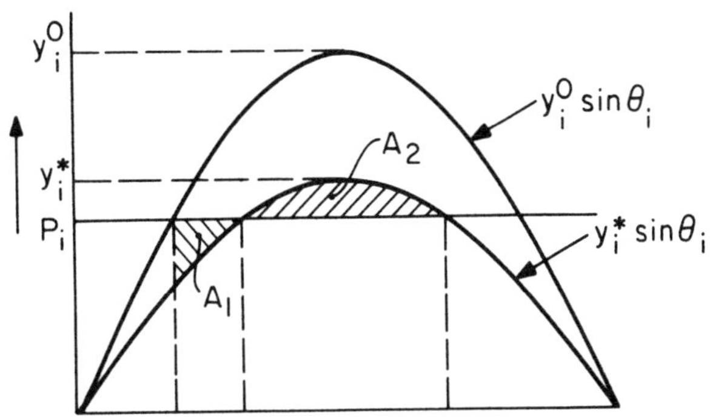

Fig. 3.3 Equal area criterion.

Stablity is lost when $A_1 > A_2$. The quantity y_i^*/y_i^o indirectly reflects the loss of admittance connected to node i. The value $(y_i^o - y_i^*)/y_i^o = z_i$ represents the relative reduction in equivalent admittance. The ratio $P_i/y_i^o = y_i$ is the injection at node i as a fraction of the steady state power limit. Equating areas A_1 and A_2 and expressing the relationship in terms of y and z gives the curve in Fig. 3.4. One can also compute a quantity called the measure of stability $s = \frac{A_2 - A_1}{A_1}$ and obtain through detailed algebraic manipulation $s = s(z_i, y_i)$. For fixed loss of admittance at node i, one can also plot s vs. y_i. For details see Siddiqee and Peschon (1969).

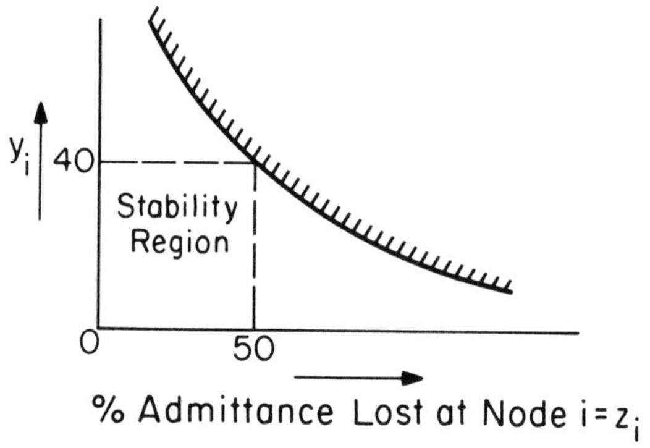

Fig. 3.4 Power limits.

Figure 3.4 can be interpreted physically as follows. If node i is connected to the remainder of the system by two lines of equal capacity and carrying the same load, the loss of one line (z = 50%) would not cause transient instability provided that the prefault injection at node i was less than 40% of the prefault transmission capacity. The methodology outlined above can be used for on-line usage also.

3.4 Cutset Energy Function

3.4.1 Mathematical model (Bergen and Hill (1981))

We consider the structure preserving model as in Sec. 2.5.1. It is a structure preserving model where all the buses in the transmission network are considered as P-V buses and Q at these nodes is not explicitly considered. Only real power loads are modeled as frequency dependent. A damping term D_i is included in the swing equation.

The equations, all in the synchronous frame of reference, are

$$M_i \ddot{\phi}_i + D_i \dot{\phi}_i = P_{mi} - \sum_{j=1}^{n+N} B_{ij} V_i V_j \sin(\phi_i - \phi_j) \qquad i = 1, 2 \ldots n \tag{3.17}$$

$$D_i\dot{\phi}_i = -P_{Di} - \sum_{j=1}^{n+N} B_{ij} |V_i||V_j| \sin(\phi_i - \phi_j) \quad i = n+1, \ldots n+N \tag{3.18}$$

we assume
$$\sum_{i=1}^{n} P_{mi} - \sum_{i=n+1}^{n+N} P_{Di} = 0 \tag{3.19}$$

Define the injected power $P_i^o = P_{mi}$ $(i = 1, 2 \ldots n)$

$= -P_{Di}$ $(i = n+1, \ldots n+N)$.

There are various ways to describe the angle variables in the network. We define the inter-nodal angles

$$\alpha_i = \phi_i - \phi_{n+N}, \quad i = 1, 2 \ldots (n+N-1)$$

and the line angle differences

$$\sigma_k = \phi_i - \phi_j, \quad k = 1, 2 \ldots \ell$$

for the kth line joining buses i and j where there are ℓ lines in the augmented network. The angle vectors $\phi = [\phi_1, \ldots \phi_{n+N}]^T$, $\alpha = [\alpha_1, \ldots \alpha_{n+N-1}]^T$ and $\sigma = [\sigma_1, \ldots \sigma_\ell]^T$ are related by the various transformations

$$\alpha = T\phi \tag{3.20}$$
$$\sigma = A^T \alpha$$

Matrix T is given by $[I_{n+N-1} \quad -e_{n+N-1}]$ and I = identity matrix and e_{n+N-1} is the n+N-1 vector with unity entries. Matrix A is the reduced incidence matrix of the augmented network. We denote the real power flow in line k as $g_k = b_k \sin \sigma_k$ where $b_k = |V_i||V_j| B_{ij} > 0$ and it is assumed that the branch k connects buses i and j. Let $P_{n+N}^o = -\sum_{i=1}^{n+N-1} P_i^o$ and let $P^o = [P_1^o, \ldots P_{n+N-1}^o]^T$. Then the load flow equations in the augmented network can be written as

$$P^o = A\, g\, (A^T\alpha) \triangleq f(\alpha)$$

Note that
$$f_i(\alpha) = \sum_{\substack{j=1 \\ \neq i}}^{n+N-1} |V_i||V_j| B_{ij} \sin(\alpha_i - \alpha_j) + b_{i,n+N} \sin \alpha_i \quad , \; i = 1,2 \ldots n+N-1 \tag{3.21}$$

It is also convenient to partition T in (3.20) as

$$T = \overset{\quad N \qquad\quad n}{\left[T_1 \quad T_2\right]} \; n+N-1$$

State space representation of Equations (3.17) and (3.18) is obtained by choosing $\omega_g = [\omega_1, \ldots \omega_n]^T$ and $\alpha = [\alpha_1, \ldots \alpha_{n+N-1}]^T$ as the state variables where $\omega_i = \dot{\phi}_i$ and α_i is as defined earlier. The result is

$$\dot{\omega}_g = -M_g^{-1} D_g \omega_g - M_g^{-1} T_2^T [f(\alpha) - P^o] \tag{3.22}$$

$$\dot{\alpha} = T_2 \omega_g - T_1 D_\ell^{-1} T_1^T [f(\alpha) - P^o] \tag{3.23}$$

where
$$M_g = \text{Diag}(M_i) \quad i = 1, 2 \ldots n$$
$$D_g = \text{Diag}(D_i) \quad i = 1, 2 \ldots n$$
$$D_\ell = \text{Diag}(D_i) \quad i = n+1, \ldots n+N$$

The equilibrium is given by $\omega_g = 0$ and α^e which is the solution of $f(\alpha^e) = P^o$. The dimension of the state space model is n+n+N-1 = 2n+N-1. There are numerous possible solutions of α^e.

3.4.2 Cutset stability criterion [Chandrashekhar and Hill (1986)]

Following a fault, a large power system tends to oscillate in coherent groups. Instability occurs when one or more of these groups (which may consist of just one machine) lose synchronism with the rest of the system. Figure 3.5 illustrates a typical view of the faulted power system. The groupings A, B, etc. are dependent on the fault. Instability occurs by the loss of synchronism across one of the cutsets C_i. Suppose that we wish to find if group A is likely to split from the system. It seems reasonable to ignore the transient energy in remote groups such as C and D. The system energy function is

$$V(\alpha, \omega_g) = \frac{1}{2} \omega_g^T M_g \omega_g + W(\alpha, \alpha^\circ) \qquad (3.24)$$

where $$W(\alpha, \alpha^\circ) = \sum_{k=1}^{\ell} b_k \, h(\sigma_k, \sigma_k^\circ)$$

$$h(\sigma_k, \sigma_k^\circ) = \int_{\sigma_k^\circ}^{\sigma_k} (\sin u - \sin \sigma_k^\circ) \, du$$

$(\alpha^\circ, 0)$ is the post-fault stable equilibrium point. If the system separation occurs across the cutset C_i, which has q lines, then the reduced energy function is

$$\frac{1}{2} \frac{M_A M_B}{M_A + M_B} (\tilde{\omega}_A - \tilde{\omega}_B)^2 + \sum_{k=1}^{q} b_k \, h(\sigma_k, \sigma_k^\circ) \qquad (3.25)$$

where $M_A = \sum_{i \varepsilon A} M_A$; $M_B = \sum_{i \varepsilon B} M_i$, $\tilde{\omega}_A$, $\tilde{\omega}_B$ are the center of speeds of groups A and B respectively. For stability this cutset energy function must be $< V_{cr}^{cutset}$ which is calculated as follows.

Let us picture the angle separation of the system across the cutset C_i as in Fig. 3.5. The function sin σ_k is shown in Fig. 3.6 and the points of intersection of this nonlinearity with sin σ_k^o identify the stable and unstable equilibrium points. In Fig. 3.6, σ_k^o is the post-fault s.e.p. and $\sigma_k^u = \Pi - \sigma_k^o$ and $\sigma_k^\ell = -\Pi - \sigma_k^o$ are the other approximate u.e.p's by acceleration and deceleration, respectively. Let μ_k^u and μ_k^ℓ be the value of the integrals defined as

$$\mu_k^u = \int_{\sigma_k^o}^{\sigma_k^u} (\sin u - \sin \sigma_k^o) \, du$$

$$\mu_k^{\ell} = \int_{\sigma_k^o}^{\sigma_k^{\ell}} (\sin u - \sin \sigma_k^o)\, du$$

Let C_i^+ be the subset of lines in set C_i agreeing with the positive orientation of C_i and C_i^- be the subset of lines having an orientation opposite to that of C_i. Then the net deceleration area across the cutset C_i is given as

$$V_{cr} = \sum_{C_i^+} b_k \mu_k^u + \sum_{C_i^-} b_k \mu_k^{\ell}$$

For implementation it is important to know the possible cutsets along which the system might separate. For instance we could consider just those cutsets containing the faulted line. The quantity V_{cr} for each of these possible "candidates" are then ranked from smallest to largest. The smallest value is then chosen as the vulnerable cutset and V_{cr} across this cutset is V_{cr}^{cutset}. To judge stability of the system, the faulted equations are integrated until $t = t_{c\ell}$ and for stability we must have the expression evaluated at $t = t_{c\ell}$ less than V_{cr}^{cutset}. In the next section we illustrate the procedure through an example.

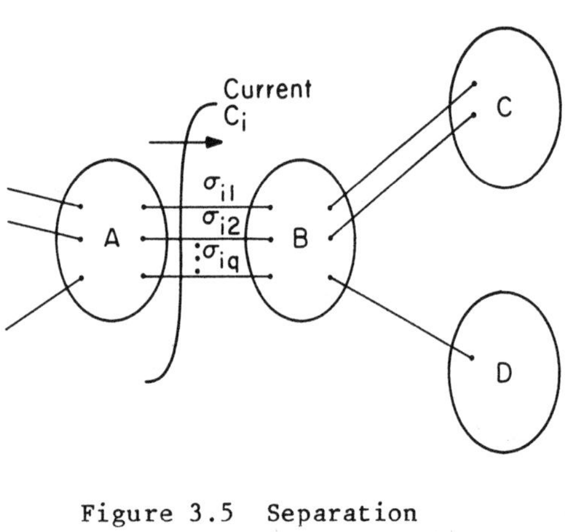

Figure 3.5 Separation across a cutset

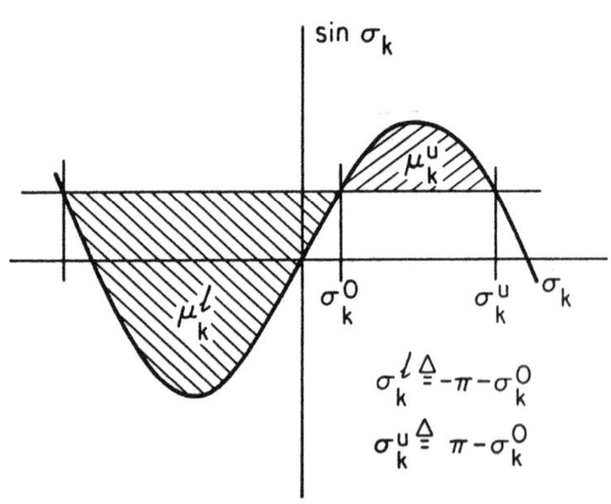

Figure 3.6 Nonlinearity

3.5 Example of cutset energy function (Chandrasekhar and Hill (1986))

Consider the two machine systems in Fig. 3.7. 1-5 are physical nodes and 6-7 are internal nodes corresponding to machines. The lines are oriented and numbered systematically as follows:

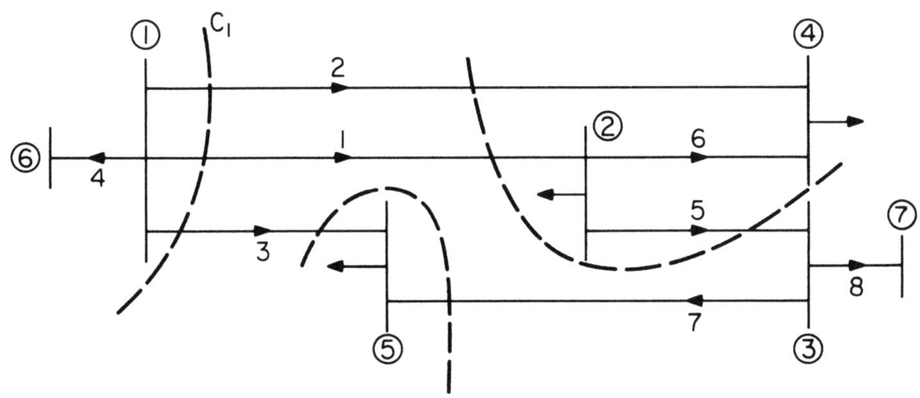

Figure 3.7 5 Bus network

TABLE 3.1

LINE DATA

Line No.	Bus no. From	Bus no. To	Angle
1	1	2	ϕ_{12}
2	1	4	ϕ_{14}
3	1	5	ϕ_{15}
4	1	6	ϕ_{16}
5	2	3	ϕ_{23}
6	2	4	ϕ_{24}
7	3	5	ϕ_{35}
8	3	7	ϕ_{37}

The matrices T and A are given by

$$T = \begin{bmatrix} 1 & 0 & 0 & 0 & 0 & 0 & -1 \\ 0 & 1 & 0 & 0 & 0 & 0 & -1 \\ 0 & 0 & 1 & 0 & 0 & 0 & -1 \\ 0 & 0 & 0 & 1 & 0 & 0 & -1 \\ 0 & 0 & 0 & 0 & 1 & 0 & -1 \\ 0 & 0 & 0 & 0 & 0 & 1 & -1 \end{bmatrix} \tag{3.26}$$

$$\begin{array}{cc} & \begin{array}{cccccccc} 1 & 2 & 3 & 4 & 5 & 6 & 7 & 8 \end{array} \quad \text{Lines} \\ A = \begin{array}{c} 1 \\ 2 \\ 3 \\ 4 \\ 5 \\ 6 \end{array} & \begin{bmatrix} 1 & 1 & 1 & 1 & 0 & 0 & 0 & 0 \\ -1 & 0 & 0 & 0 & 1 & 1 & 0 & 0 \\ 0 & 0 & 0 & 0 & -1 & 0 & 1 & 1 \\ 0 & -1 & 0 & 0 & 0 & -1 & 0 & 0 \\ 0 & 0 & -1 & 0 & 0 & 0 & -1 & 0 \\ 0 & 0 & 0 & -1 & 0 & 0 & 0 & 0 \end{bmatrix} \quad \text{Nodes} \end{array} \tag{3.27}$$

Node 7, corresponding to the internal node of the swing bus machine, is omitted in (3.27) since it is a reduced incidence matrix.

Denoting $\phi_i - \phi_j = \phi_{ij}$, we verify from the above matrices and using (3.16) that

$$\begin{bmatrix} \alpha_1 \\ \alpha_2 \\ \alpha_3 \\ \alpha_4 \\ \alpha_5 \\ \alpha_6 \end{bmatrix} = \begin{bmatrix} \phi_{17} \\ \phi_{27} \\ \phi_{37} \\ \phi_{47} \\ \phi_{57} \\ \phi_{67} \end{bmatrix} \quad \text{and} \quad \begin{bmatrix} \sigma_1 \\ \sigma_2 \\ \sigma_3 \\ \sigma_4 \\ \sigma_5 \\ \sigma_6 \\ \sigma_7 \\ \sigma_8 \end{bmatrix} = \begin{bmatrix} \phi_{12} \\ \phi_{14} \\ \phi_{15} \\ \phi_{16} \\ \phi_{23} \\ \phi_{24} \\ \phi_{35} \\ \phi_{37} \end{bmatrix}$$

The alternative way of partitioning T is $[T_1 \quad T_2]$ where

$$T_1 = \overset{5}{\begin{bmatrix} 1 & 0 & 0 & 0 & 0 \\ 0 & 1 & 0 & 0 & 0 \\ 0 & 0 & 1 & 0 & 0 \\ 0 & 0 & 0 & 1 & 0 \\ 0 & 0 & 0 & 0 & 1 \\ 0 & 0 & 0 & 0 & 0 \end{bmatrix}} \qquad T_2 = \overset{2}{\begin{bmatrix} 0 & -1 \\ 0 & -1 \\ 0 & -1 \\ 0 & -1 \\ 0 & -1 \\ 1 & -1 \end{bmatrix}} \tag{3.28}$$

$g(\sigma)$ and $f(\alpha)$ are given by

$$g(\sigma) = \begin{bmatrix} b_1 \sin \phi_{12} \\ b_2 \sin \phi_{14} \\ b_3 \sin \phi_{15} \\ b_4 \sin \phi_{16} \\ b_5 \sin \phi_{23} \\ b_6 \sin \phi_{24} \\ b_7 \sin \phi_{35} \\ b_8 \sin \phi_{37} \end{bmatrix}, \quad f(\alpha) = \begin{bmatrix} g_1(\sigma_1) + g_2(\sigma_2) + g_3(\sigma_3) + g_4(\sigma_4) \\ -g_1(\sigma_1) + g_5(\sigma_5) + g_6(\sigma_6) \\ -g_5(\sigma_5) + g_7(\sigma_7) + g_8(\sigma_8) \\ -g_2(\sigma_2) - g_6(\sigma_6) \\ -g_3(\sigma_3) - g_7(\sigma_7) \\ -g_4(\sigma_4) \end{bmatrix} \tag{3.29}$$

Substituting the various matrices in (3.22) and (3.23) gives

$$\begin{aligned}
M_6 \dot{\omega}_6 + D_6 \omega_6 &= P_6^o - b_4 \sin \phi_{61} \\
&= P_{m1} - |V_1||V_6| B_{16} \sin \phi_{61} \\
M_7 \dot{\omega}_7 + D_7 \omega_7 &= P_7^o - b_8 \sin \phi_{73} \\
&= P_{m2} - |V_3||V_7| B_{37} \sin \phi_{73} \\
D_1 \dot{\phi}_{17} &= -D_1 \omega_7 - \left(g_1(\sigma_1) + g2(\sigma 2) + g_3(\sigma_3) + g_4(\sigma_4)\right) + P_1^o \\
D_2 \dot{\phi}_{27} &= -D_2 \omega_7 - \left(-g_1(\sigma_1) + g_5(\sigma_5) + g_6(\sigma_6)\right) + P_2^o \\
D_3 \dot{\phi}_{37} &= -D_3 \omega_7 - \left(-g_5(\sigma_5) + g_7(\sigma_7) + g_8(\sigma_8)\right) + P_3^o \\
D_4 \dot{\phi}_{47} &= -D_4 \omega_7 - \left(-g_2(\sigma_2) - g_6(\sigma_6)\right) + P_4^o \\
D_5 \dot{\phi}_{57} &= -D_5 \omega_7 - \left(-g_3(\sigma_3) - g_7(\sigma_7)\right) + P_5^o \\
D_6 \dot{\phi}_{67} &= D_6(\omega_6 - \omega_7)
\end{aligned} \tag{3.30}$$

It can be easily verified that Eq. (3.30) is equivalent to (3.17) and (3.18) namely

$$M_6 \ddot{\phi}_6 + D_6 \dot{\phi}_6 = P_{m1} - |V_1||V_6| B_{16} \sin \phi_{61}$$

$$M_7 \ddot{\phi}_7 + D_7 \dot{\phi}_7 = P_{m2} - |V_3||V_7| B_{37} \sin \phi_{73}$$

$$D_i \dot{\phi}_i = P_i^o - \sum_{j=1}^{7} |V_i||V_j| \sin \phi_{ij} \qquad i = 1,2 \ldots 5 \tag{3.31}$$

Equations (3.30) are in the minimal state space which is 2n+N-1 = 8.

Stability assessment algorithm

The basic steps are as follows:

1. For a given fault and anticipated protection actions, find the post-fault stable equilibrium point.

2. Tabulate the vulnerability indices V_{cr} for the post-fault network cutsets and for a given fault select the one across which separation is likely to occur. Since the number of cutsets may be large, the "candidate" cutsets must be chosen carefully. For instance, one could consider just those cutsets containing the faulted line. The V_{cr} for these cutsets is then ranked from smallest to the largest. The cutsets corresponding to the smaller vulnerability indices are the weaker links in the system for the given fault. Generally the cutset corresponding to the smallest V_{cr} is chosen as V_{cr}^{cutset}. Sometimes other information can be used to override this procedure.

3. For assessing stability, integrate the faulted system until $t = t_{c\ell}$. If at $t_{c\ell}$, $V(\alpha, \omega_g) < V_{cr}^{cutset}$, then the system is stable. To compute t_{cr}, it is the time when $V(\alpha, \omega_g) = V_{cr}^{cutset}$ along the faulted trajectory.

Numerical results

The line data, load flow data, branch vulnerability coefficients and the cutset vulnerability indices are given below for the two machine system of Fig. 3.7 (H_6 = 50 sec, H_7 = 10 sec)

TABLE 3.2 LINE DATA

Line No.	R(p•u)	x(p•u)
1	0.10	0.40
2	0.15	0.10
3	0.05	0.20
4	0.0	0.25
5	0.05	0.20
6	0.10	0.40
7	0.05	0.20
8	0.0	1.50

TABLE 3.3 LOAD FLOW DATA

Bus No.	V(p•u)	δ(degrees)	P(p•u)	Q(p.u)
1	1.02	0.0	0.954	0.27
2	0.955	-6.32	-0.6	-0.3
3	1.04	-1.98	0.697	-0.527
4	0.923	-9.4	-0.4	-0.1
5	0.993	-4.08	-0.6	-0.2

P and Q are injected powers.

TABLE 3.4 BRANCH VULNERABILITY COEFFICIENTS

Line No.	σ_k^o (rad)	b_k	μ_k^μ	μ_k^ℓ
1	0.1103	2.29	1.67	2.36
2	0.1641	1.48	1.51	2.54
3	0.0712	4.77	1.78	2.23
4	0.2047	4.53	1.40	2.68
5	0.0757	4.67	2.24	1.77
6	0.0538	2.07	1.83	2.17
7	0.0367	4.86	1.89	2.12
8	0.5093	1.43	3.77	0.71

TABLE 3.5 CUTSET VULNERABILITY INDICES

Cutset	Lines	V_{cr}
1	1,2,3	14.55
2	1,3,6	16.82
3	1,2,7	16.36
4	2,3,5,6	25.72
5	3,7	17.66
6	2,5,6,7	26.81
7	5,7	17.43
8	1,5,6	16.58
9	1,6,7	17.49
10	3,5	18.97
11	2,6	6.04
12	1,2,5	14.32
13	8	5.39

Cutset references are as for the first line in the cutset. We consider the fault at buses 1, 3, and 5. These are faults which are self-clearing, i.e., the post-fault configuration is the same as the pre-fault configuration.

Fault on bus 1

For this fault we might expect splitting on cutsets 1, 3, or 5. Cutset 1 is most vulnerable and hence $V_{cr}^{cutset} = 14.55$ for this fault.

Fault on bus 3

For this fault we consider cutsets 5, 7, 8 and 12. Cutset 12 is most vulnerable and hence $V_{cr}^{cutset} = 14.32$ for this fault.

Fault on bus 5

For this fault although cutset 1 gives the minimum value of V_{cr}, cutset 5 is likely to be the vulnerable one since it connects two generators. Hence, $V_{cr}^{cutset} = 17.66$. Table 3.6 gives the critical clearing times for the various faults and these are compared with the actual t_{cr} and the t_{cr} obtained using the system energy function. The results are quite satisfactory.

TABLE 3.6 CRITICAL CLEARING TIMES FOR 2-MACHINE SYSTEM

Faulted bus	Cutset of separation	Critical clearing time secs.		
		Step-by step	Total energy function	Cutset energy function
1	1	0.39-0.40	0.30-0.31	0.39-0.40
3	12	0.40-0.41	0.22-0.23	0.39-0.40
5	5	0.34-0.35	0.22-0.23	0.33-0.34

3.6 Extended Equal Area Criterion (EEAC) Method [Xue et al. (1988)]

In this method, the multi-machine system is decomposed into two subsets; one consisting of the "cluster of critical machines" (called the "critical cluster") and the other comprising the remaining system. Each group is reduced to a single machine represented by its COI. Then the two-machine equivalent is reduced to a single-machine infinite bus system.

Consider the multi-machine model

$$\dot{\delta}_i = \omega_i \qquad i=1,\ 2,\ \ldots\ n$$

$$M_i\dot{\omega}_i = P_i - \sum_{\substack{j=1 \\ \neq i}}^{n} E_i E_j Y_{ij} \cos(\delta_i - \delta_j - \theta_{ij}) \tag{3.32}$$

For a given disturbance the procedure consists of (i) drawing up a list of "candidate" critical clusters by using the generator's initial accelerations to "prefilter" the plausible candidates, (ii) developing a one machine infinite bus equivalent, and (iii) computing by turn the critical clearing times for each "candidate" critical cluster. The actual critical cluster will be the one yielding the smallest among the above values.

The single machine infinite bus system, after aggregation, etc., is described by

$$M_{eq}\ddot{\delta} = P_m - [P_c + P_{max}\sin(\delta-\nu)] \tag{3.33}$$

With a different notation it is similar to Eq. (1.38). δ is the angle between the COI of the "candidate" cluster and the COI of the rest of the system. In terms of notation, (see sec. 1.5).

$$P_c = P_e', \ P_{max} = P_e^{max}, \ \nu = \gamma$$

The parameters P_c and ν assume different values during faulted and post-fault periods. In Fig. 3.8, δ_o is the pre-fault angle. The post fault stable equilibrium point is δ_s. The accelerating and decelerating areas are given by

$$A_{acc} = (P_m - P_{cD})(\delta_c - \delta_o) + P_{max,D}\left[\cos(\delta_c - \nu_D) - \cos(\delta_o - \nu_D)\right] \qquad (3.34)$$

$$\begin{aligned} A_{dec} &= (P_{cP} - P_m)(\pi - \delta_c - \delta_s + 2\nu_p) \\ &\quad + P_{max,P}\left[\cos(\delta_c - \nu_p) + \cos(\delta_s - \nu_p)\right] \end{aligned} \qquad (3.35)$$

The transient stability margin is given by $\eta = A_{dec} - A_{acc}$ for a given t and corresponding δ_c. The critical clearing time is when $\eta = 0$. For a given disturbance and its corresponding critical cluster (i) compute δ_c by solving the equation $\eta = 0$ and (ii) compute t_{cr} from

$$\delta_c = \delta_o + \alpha_1^{-1}\alpha_2^{-2}\gamma\frac{t_{cr}^2}{2} + \alpha_1^{-1}\alpha_2^{-4}\gamma^{(2)}\frac{t_{cr}^4}{24} \qquad (3.36)$$

where γ and $\gamma^{(2)}$ denote the acceleration and its second derivative in the SMIB system, at $t = o^+$, i.e., immediately after the disturbance. α_1 and α_2 are correction factors to improve the trajectory approximation by Taylor Series.

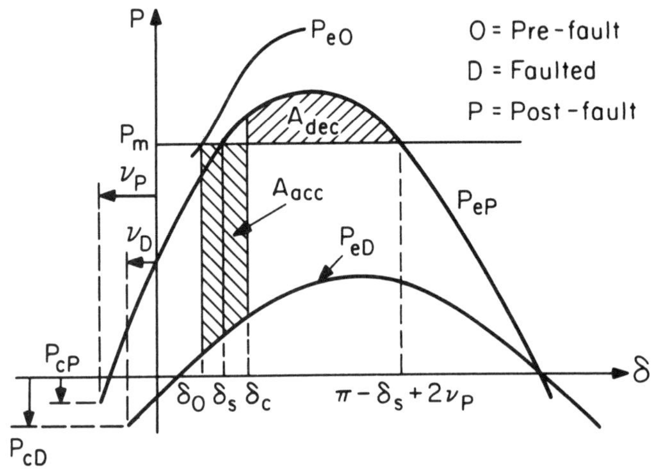

Figure 3.8 The energy balance picture.

Instead of equal area criterion (EAC), we can also use the energy function as in Sec. 1.5.3. But EAC has more appeal to practicing engineers. Since this approach has been successfully tested on a large number of systems, we will discuss it also in Chapter VI on practical applications.

3.7 The Quasi Unstable Equilibrium Point (QUEP) Method [Lemmon et al. (1988)]

Conceptually this method has a framework similar to that in the previous section but it differs in the technique of clustering depending on the fault. Instead of supplying the "candidate" cluster it picks the "critical" cluster on-line. Also, unlike in the previous method, t_{cr} is not computed but stability assessment is made given a $t_{c\ell}$. At clearing, the clustering is done primarily on the basis of rotor speed values followed by rotor angle separations within each cluster. There may be several candidate clusters out of which the one having the least stability margin is the final cluster. The machines in this final cluster are aggregated into a single machine and the rest of the machines are similarly combined into one machine. Then the theory of single machine infinite bus is applicable.

3.8 Decomposition-Aggregation Method

The concept of the decomposition-aggregation method for stability analysis can be carried out when the system separates into more than two parts. It was proposed by Nityanandan and Nagappan (1979) in the following manner.

Based on the accelerations of the machines at $t = 0^+$ relative to the inertial accelerations, the machines are grouped in three categories:

1) Machines with large relative acceleration

2) Machines with little acceleration or deceleration

3) Machines with large relative deceleration

Thus, we get three groups of machines. Each group can be combined into an equivalent machine resulting in a 3-machine system. Sometimes, depending on the nature of the fault, one may not have either the accelerating or the decelerating machines so that we have a 2-machine equivalent which has been discussed earlier. Nityanandan (1979) illustrates the method on a 7-machine system. Rahimi and Schaffer (1987) use a 2-machine equivalent with time varying parameters, which will be discussed in Chapter VI along with the work of Xue et al. (1988). We next propose a generalization of these concepts based on time-scales and singular perturbation theory.

3.9 Time-Scale Energies

This section is based on the work of Othman (1987). When a system becomes unstable, it is assumed that it separates into r ($r \geq 2$) groups of machines.

Each group has a coherent set of machines. The intermachine oscillations in each group do not contribute to system instability. However, the motions of the COI of each group are of interest in determining the energy function, critical energy, etc. In reality, the coherent set of machines is determined by the particular fault. It is easier, however, to determine the coherent areas based on pre-fault state and linearization. It is known as the 'slow coherency' approach due to Kokotovic et al. (1982) and can be justified via singular perturbations. We will assume this to be true for the rest of the presentation.

3.9.1 Mathematical model and time-scale energy functions

Let δ_i be the angle of the ith machine with respect to the synchronous frame of reference. Let the system be divided into r coherent areas where each machine in the system is assigned to an area. The set containing the indices of the machines in area α is denoted by J_α. The center of inertia (COI) angle for area α can be written

$$\begin{aligned} y_\alpha &= \sum_i^\alpha M_i \delta_i / \sum_i^\alpha M_i \\ &= \sum_i^\alpha M_i \delta_i / M_\alpha \end{aligned} \tag{3.37}$$

where $M_\alpha = \sum_i^\alpha M_i$

$\sum_i^\alpha$ = summation over all i in J_α

Also define $\tilde{y}_\alpha = y_\alpha - \delta_0$ where δ_0 is the global COI angle. The rotor angle of machine i in area α with respect to its local COI is denoted by $\tilde{x}_i = \delta_i - y_\alpha$.

It is straightforward to show that

$$\delta_i - \delta_j = \delta_{ij} = \tilde{x}_{ij} \qquad \text{if } i \text{ and } j \text{ are in the same area}$$

$$\delta_{ij} = \tilde{x}_{ij} + y_{\alpha\beta} \qquad \text{if } i \text{ is in area } \alpha \text{ and } j \text{ is in area } \beta \tag{3.38}$$

y_α and $\tilde{x}_{ij}$ are the slow and fast variables, respectively, as identified from the slow coherency approach [Chow, (1982)].

The swing equations of a multi-machine power system are given by

$$M_i \frac{d^2\delta_i}{dt^2} = P_{m_i} - \sum_j^{\alpha} [V_i V_j B_{ij} \sin\delta_{ij} + V_i V_j G_{ij} \cos\delta_{ij}]$$
$$- \sum_{\beta\neq\alpha}^{r} \sum_j^{\beta} [V_i V_j B_{ij} \sin \delta_{ij} + V_i V_j G_{ij} \cos \delta_{ij}]$$
$$\text{for all } i \in J_\alpha$$
$$\alpha = 1, 2, \ldots, r \tag{3.39}$$

where M_i, P_{mi}, V_i, B_{ij}, G_{ij} have the usual meaning.

Slow Global Subsystems

Using equations (3.37) and (3.38), and substituting into equation (3.39), we obtain the following equation for area α.

$$M_\alpha \frac{d^2 y_\alpha}{dt^2} = \sum_i^{\alpha} P_{m_i} - \sum_i^{\alpha} \sum_j^{\alpha} V_i V_j G_{ij} \cos \tilde{x}_{ij}$$
$$- \sum_{\beta\neq\alpha}^{r} \sum_i^{\alpha} \sum_j^{\beta} [V_i V_j B_{ij} \sin (y_{\alpha\beta} + \tilde{x}_{ij})$$
$$+ V_i V_j G_{ij} \cos (y_{\alpha\beta} + \tilde{x}_{ij})] \tag{3.40}$$
$$\alpha=1, 2, \ldots r$$

If the machines i and j are both in area α, then $\tilde{x}_{ij}$ is just the angular difference within an area. If α is the study area, this quantity is predominantly a fast oscillatory part superimposed on a small slowly varying part. In the slow time scale we average the fast oscillatory part [Chow, Allemong and Kokotovic, (1978)] from which $\tilde{x}_{ij} \simeq \tilde{x}^s_{ij}$ where $\tilde{x}^s_{ij}$ is the slow component of $\tilde{x}_{ij}$. If α is not the study area, $\tilde{x}_{ij}$ can still be viewed as being made up of a fast and slow part, but both components are small. If machines i and j are in diferent areas and one of them is in the disturbed area, then $\tilde{x}_{ij}$ is a predominantly fast variable superimposed on a slowly varying part. The use of averaging will again result in $\tilde{x}_{ij} \simeq \tilde{x}^s_{ij}$. When machines i and j are in different nondisturbed areas, the slow and fast components of $\tilde{x}_{ij}$ will be small and $\tilde{x}_{ij} \simeq \tilde{x}^s_{ij}$. As a further simplification, we let $\tilde{x}^s_{ij}$ be equal to the post-fault equilibrium value $\tilde{x}^0_{ij}$. We can then rewrite subsystem (3.40) as

$$M_\alpha \frac{dy_\alpha^2}{dt^2} = P_\alpha - \sum_{\beta\neq\alpha}^{r} \sum_{i}^{\alpha} \sum_{j}^{\beta} [V_i V_j B_{ij} \sin (y_{\alpha\beta} + \tilde{x}^0_{ij}) + V_i V_j G_{ij} \cos (y_{\alpha\beta} + \tilde{x}^0_{ij})] \tag{3.41}$$

where

$$P_\alpha = \sum_{i}^{\alpha} P_{m_i} - \sum_{i}^{\alpha} \sum_{j\neq i}^{\alpha} V_i V_j G_{ij} \cos \tilde{x}^0_{ij}.$$

The energy function associated with subsystem (3.41) is called V_{slow} and is a function of slow variables only.

Fast Local Subsystems

In the fast time scale, the slow variables are set at their initial conditions. Using (3.39) we get

$$M_i \frac{d^2\tilde{x}_i}{dt^2} = P_{mi} - \sum_j^{\alpha} [V_i V_j B_{ij} \sin \tilde{x}_{ij}$$

$$+ V_i V_j G_{ij} \cos \tilde{x}_{ij}]$$

$$- \sum_{\beta \neq \alpha}^{r} \sum_j^{\beta} \left(V_i V_j B_{ij} \sin (y_{\alpha\beta}^0 + \tilde{x}_{ij})\right.$$

$$\left. + V_i V_j G_{ij} \cos (\tilde{y}_{\alpha\beta}^0 + \tilde{x}_{ij})\right)$$

$$i = 1, 2, \ldots, n_\alpha$$

$$\alpha = 1, 2, \ldots, r. \qquad (3.42)$$

Since the connections between the areas are weak (i.e., G_{ij}, $B_{ij} = O(\varepsilon)$), the third term on the right-hand side of (3.42) is generally set to a constant. However, we represent this term as equivalent injections at the boundary buses to obtain a physically reduced-order model of the area. To give a physical meaning to the two models, subsystem (3.41) represents an r machine system whose individual inertias are the sum of the inertias of each area. It has (r-1) slow oscillatory modes. Subsystem (3.42) represents for each area α, an n_α machine subsystem, each subsystem having $(n_\alpha - 1)$ fast oscillatory modes. Equations (3.41) and (3.42) are decoupled (Fig. 3.9).

Note that we have retained the physical picture throughout in deriving (3.41) and (3.42) which could be derived formally using the singular perturbations approach [Chow, (1982)].

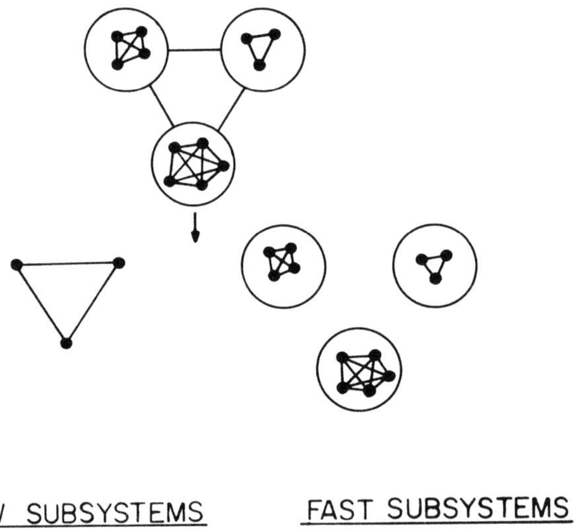

Slow Subsystems   Fast Subsystems

Fig. 3.9. Decomposition into slow and fast subsystems.

Using the subsystems (3.41) and (3.42), we can decompose the system transient energy function into a slow component with r machines and r fast components, each with n_α machines. In order to explain the nature of slow and fast energies and the role they play in system stability, we take a three-machine example (Fig. 3.10) where machines 2 and 3 are in one area (area 1) and machine 1 is by itself (area 2). While area 1 has fast dynamics, area 2 does not. The single line diagram of the system is given in Fig. 3.10. Note that weak connections exist between the two areas because they are connected by two relatively high impedance transmission lines. The load flow data and relevant parameters are given in Table 3.7. The eigenvalues of the linearized model about the equilibrium point are 0, o, $\pm j3.6163$ and $\pm j24.307$. The slow model has $\pm j3.6163$ and the fast mode has $\pm j24.307$ as the eigenvalue. The area grouping is machines

2 and 3 in area one and machine 1 by itself as area two. The value of ε representing the time-scale separation if the equations are put in a singularly perturbed form is $\left(\frac{3.6363}{24.307}\right)^2 = 0.022$. The energy functions for the two sub-systems of the three-machine system are V_{slow} and $V_{fast,\alpha}$. For the slow sub-system, with the subscripts KE denoting kinetic energy and PE denoting potential energy, we have

$$V_{slow} = V^{KE}_{slow} + V^{PE}_{slow}$$

where

$$V^{KE}_{slow} = \frac{1}{2} M_1 \dot{\tilde{y}}_1^2 + \frac{1}{2} (M_2 + M_3) \dot{\tilde{y}}_2^2 \tag{3.43}$$

$$\begin{aligned} V^{PE}_{slow} = &- P_1(\tilde{y}_1 - \tilde{y}_1^0) - P_2(y_2 - \tilde{y}_2^0) \\ &- (C_{12} + C_{13}) [\cos (\tilde{y}_1 - \tilde{y}_2) \\ &- \cos (\tilde{y}_1^0 - \tilde{y}_2^0)] \\ &+ (D_{12} + D_{13}) \int_{\tilde{y}_1^0+\tilde{y}_2^0}^{\tilde{y}_1+\tilde{y}_2} \cos (\tilde{y}_1-\tilde{y}_2) d(\tilde{y}_1+\tilde{y}_2) \end{aligned} \tag{3.44}$$

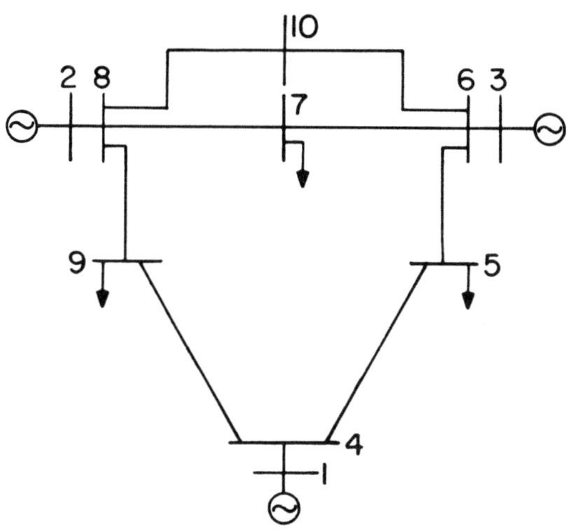

Fig. 3.10 Three-machine system.

Similarly, for the fast subsystems,

$$V_{fast,1} = V_{fast,1}^{KE} + V_{fast,1}^{PE}$$

where $$V_{fast,1}^{KE} = \frac{1}{2} M_2 \dot{\tilde{x}}_2^2 + \frac{1}{2} M_3 \dot{\tilde{x}}_3^2 \tag{3.45}$$

$$\begin{aligned} V_{fast,1}^{PE} = & - P_2(\tilde{x}_2 - \tilde{x}_2^0) - P_3(\tilde{x}_3 - \tilde{x}_3^0) \\ & - C_{23} \left[\cos(\tilde{x}_2 - \tilde{x}_3) - \cos(\tilde{x}_2^0 - \tilde{x}_2^0)\right] \\ & + D_{23} \int_{\tilde{x}_2^0+\tilde{x}_3^0}^{\tilde{x}_2+\tilde{x}_3} \cos(\tilde{x}_2 - \tilde{x}_3)\, d(\tilde{x}_2 + \tilde{x}_3) \end{aligned}$$

$$V_{fast,2} \equiv 0 \tag{3.46}$$

TABLE 3.7

Machine Parameters

Generator	X'_d (Pu)	H(secs)
1	0.06080	23.5625
2	0.01198	6.4090
3	0.01813	3.01600

Load Flow Data

Bus	V (p.u.)	θ(deg)	Generation (p.u.)		Load (p.u.)	
			P	Q	P	Q
1	1.0400	0.0000	0.5295	-1.9433	0.0000	0.0000
2	1.02500	0.82715	1.5000	-0.41891	0.0000	0.0000
3	1.02500	-0.32612	1.2000	-0.51863	0.0000	0.0000
4	1.14800	-1.4639	0.0000	0.0000	0.0000	0.0000
5	1.13130	-10.8832	0.0000	0.0000	0.9000	0.5000
6	1.05688	-4.04797	0.0000	0.0000	0.0000	0.0000
7	1.04807	-4.79512	0.0000	0.0000	1.0000	0.5000
8	1.05452	-4.14866	0.0000	0.0000	0.0000	0.0000
9	1.11067	-12.759	0.0000	0.0000	1.2500	0.6000
10	1.05638	-4.112	0.0000	0.0000	0.0000	0.0000

Line Parameters

Line	From	To	Resistance (Pu)	Reactance (Pu)	Total Line Charging (Pu)
1	1	4	0.0000	0.0576	0.0000
2	2	8	0.0000	0.0625	0.0000
3	3	6	0.0000	0.0586	0.0000
4	4	5	0.1700	0.9200	1.5800
5	4	9	0.1000	0.8500	1.7600
6	5	6	0.0390	0.1700	0.3580
7	6	7	0.0039	0.0336	0.0696
8	7	8	0.0028	0.0240	0.0496
9	8	9	0.0320	0.1610	0.3060
10	8	10	0.0028	0.0240	0.0496
11	6	10	0.0039	0.0336	0.0696

$C_{ij} = V_i V_j B_{ij}$ and $D_{ij} = V_i V_j G_{ij}$ where G_{ij} and B_{ij} are the transfer conductances and susceptances between machines i and j. For the area 1 (fast area), before obtaining G_{23} and B_{23}, at the boundary buses connecting areas 1 and 2, an equivalent injection corresponding to tie line powers is introduced to compensate for the neglect of the third term on the R.H.S. of Eq. (3.42).

Depending on the disturbance, three possibilities exist:

(1) Excite the slow dynamics predominantly,

(2) Excite the fast dynamics predominantly,

(3) Excite both the slow and fast dynamics.

We discuss qualitatively and quantitatively one of the cases for the three-machine, system in the next section.

3.9.2 Numerical example

The details of the three-machine system are given in Table 3.7. Three three-phase fault cases were considered, i.e. those at buses 7, 10 and 8, with no line switching. Each resulted in the two areas separating from each other. For a three-phase fault at bus 7 which is in area 1, the following plots are shown for the fault-on trajectories (Fig. 3.11).

(i) K.E. for V_{syst}, V_{slow}, $V_{fast,1}$ and $V_{slow} + V_{fast,1}$

(ii) P.E. for V_{syst}, V_{slow}, $V_{fast,1}$ and $V_{slow} + V_{fast,1}$

(iii) K.E. + P.E. for V_{syst}, V_{slow}, $V_{fast,1}$ and $V_{slow} + V_{fast,1}$

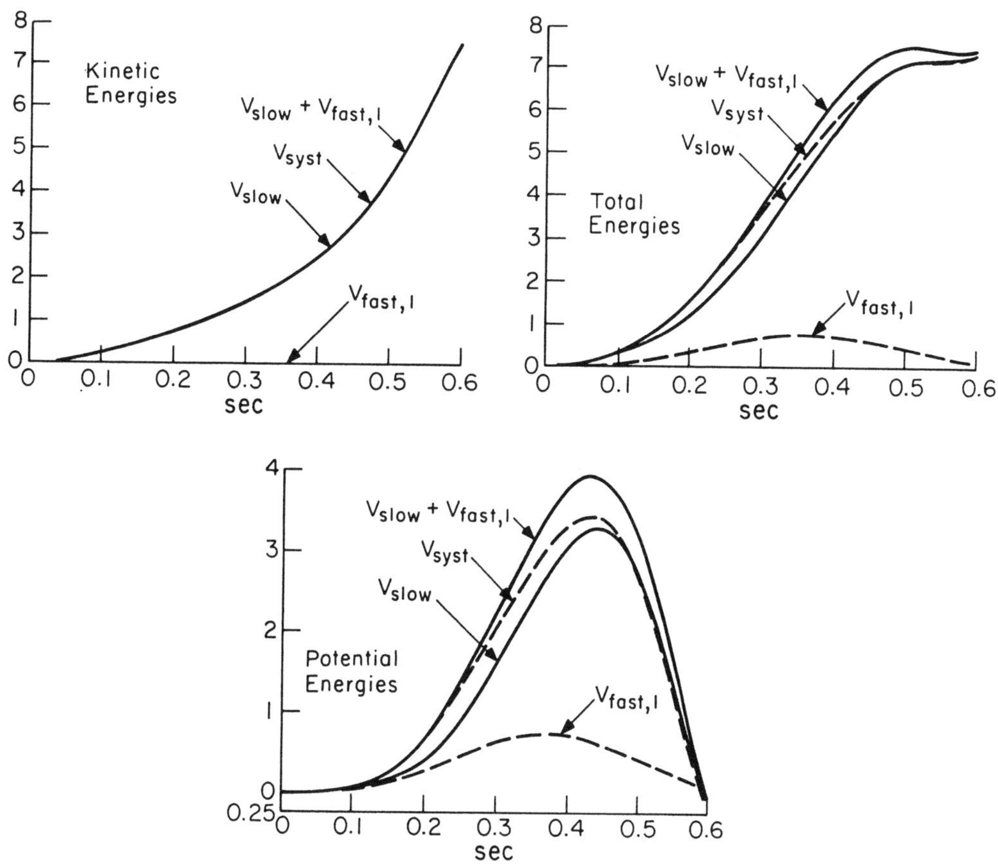

Fig. 3.11 Fast and slow energies in a 3-machine system.

To evaluate the critical clearing times for the two decoupled subsystems, we use any technique available. We used the potential energy boundary surface (PEBS) method. For these disturbances, the critical clearing times for the fast subsystem were always longer than those for the slow subsystem. Thus, the limiting clearing times are those given by the slow subsystem. The critical clearing time estimates given by the slow subsystem compare quite favorably to the critical clearing times obtained using step-by-step integration (see Table

3.8). The critical clearing times obtained from the total system (3 machine) energy functions are also given in Table 3.8. Note that in all three cases the slow subsystem energy functions produced more accurate results. The slow mode gets excited more than the fast mode and results in an inter-area instability. The rotor angles of the machines for t_{cr}^- and t_{cr}^+ are shown in Fig. 3.12 for fault at bus 7. Note the intermachine oscillations in area 1 between machines 2 and 3. In this case the slow energy is also the energy associated with system separation [Fouad and Vittal, (1983)]. However, the formulation of slow and fast energy function is more general when the mode of instability is complex.

TABLE 3.8 Critical Clearing Times Obtained by Step-by-Step, System Energy and Slow Energies

Fault at Bus #	t_{cr} (secs) step-by-step	System Energy V_P^{max}	System Energy t_{cr} (secs)	Slow Energy V_P^{max}	Slow Energy t_{cr} (secs)
7	0.307	3.431	0.2975	3.293	0.3100
10	0.316	3.403	0.3055	3.292	0.3185
8	0.250	3.303	0.249	3.293	0.250

If the coherent areas are determined based on "critical generators" or "clustering" based on the actual fault location, then the methodology becomes somewhat similar to those discussed earlier but in more general terms. The time scale energy can be further modified by taking into account the effect of the fast dynamics on the slow dynamics through a decoupling methodology. For details see Othman and Pai (1988).

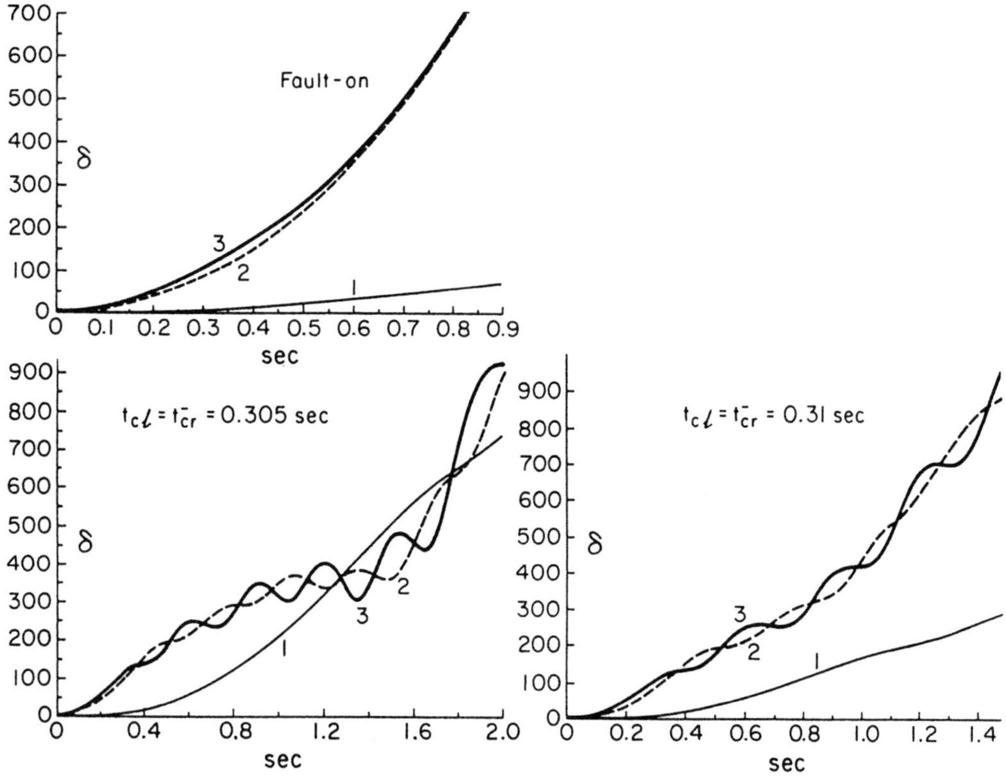

Fig. 3.12 Rotor angles of machines for t_{cr} = .305 and t_{cr} = .31 secs for a fault at bus 7

3.10 Conclusions

In this chapter we have dealt with a few of the reduced order energy functions which can form the basis of on-line dynamic security assessment. These approaches are in the same category as the familiar equal area criterion. Time-scale energies have also been discussed. Some of the techniques which have been tried on realistic systems will be discussed in Chapter VI.

Chapter 4

ENERGY FUNCTIONS WITH DETAILED MODELS OF SYNCHRONOUS MACHINES AND ITS CONTROL

4.1 Introduction

So far we have discussed energy functions with classical models for both the internal node representation and the structure preserving framework. It is well known that although the power system may be stable in the first swing, for which the classical model is adequate, it may become unstable in subsequent swings. In such cases it is necessary to consider detailed generator modeling, excitation system modeling, etc. Early work in this area was done by Siddiqee [1968] for single machine systems, and Rai [1973] and Kakimoto et al. [1979, 1980] for multimachine systems. Rai considered the multivariable Popov criterion while Kakimoto used the Moore Anderson theorem and both of them used the flux decay model. Sasaki [1979] considered the flux decay model where E'_q was considered as a parameter variation. Athay et al. [1979] extended this approach to multimachine systems. Tsolas et al. [1985] considered the flux decay model with a structure preserving network while Padiyar and Sastry [1987] considered the structure preserving model along with the excitation system modeling. Bergen, Hill and DeMarco [1984] mathematically justified the existence of a Lyapunov function for structure preserving model with flux decay and nonlinear loads. Miyagi and Bergen [1986] extended the approach of Kakimoto et al. [1980] to include automatic voltage regulators. Fouad et al. [1988] considered the effect of excitation control on the first swing stability by a parametrization of the internal node voltages analogous to Sasaki [1979] but with a different set of approximations. Sauer et al. [1988] considered a two axis model and the excitation system along with structure preserving framework but constant impedance loads. The work of Behera [1988] contains additional results in terms of nonlinear loads and sensitivity analysis using a numerical Lyapunov function. Tsolas et al. [1985], Padiyar and Sastry [1987] and Padiyar and Ghosh [1989] considered quasi-analytical energy functions for detailed models. Numerical investigations into the effects of fast exciter models and HVDC links on realistic power system models have been reported by Ni and Fouad [1987]. In this chapter we review some of the above approaches. In some instances the details are omitted but the general framework is indicated.

4.2 <u>Single Machine System with Flux Decay Model</u> [Siddiqee (1968), Willems (1969)]

The equations of the post-fault system are

$$M \frac{d^2\delta}{dt^2} = P_i - \frac{E'_q E}{(x_{12}+x'_d)} \sin \delta \tag{4.1}$$

$$\frac{dE'_q}{dt} = - \frac{1}{T'_{do}} E'_q - \frac{(x_d - x'_d)}{T'_{do}} I_d + \frac{1}{T'_{do}} E_{fd} \tag{4.2}$$

where
$$I_d = \frac{E'_q}{(x_{12}+x'_d)} - \frac{E}{x_{12}} \cos \delta$$

where x_{12} is the reactance between the generator terminal and the infinite bus whose voltage magnitude is E and phase angle is zero. E'_q is the voltage proportional to the field flux. The stable equilibrium point is given by $\dot{\delta} = 0$, $\delta = \delta^s$ and $E'_q = e$. It also follows that $P_i = \frac{Ee \sin\delta^s}{(x_{12}+x'_d)}$. It is obtained by setting the derivates equal to zero in (4.1) and (4.2) and solving for δ and E'_q. The equilibrium point is then shifted to the origin by defining new state variables as

$$\begin{aligned} x_1 &= \delta - \delta^s \\ x_2 &= \dot{\delta} \\ x_3 &= E'_q - e \end{aligned} \tag{4.3}$$

The state space equations (4.1) and (4.2) now become

$$\dot{x}_1 = x_2 \tag{4.4}$$

$$\dot{x}_2 = \frac{P_i}{M} - \frac{K_1}{M}(x_3 + e)\sin(x_1 + \delta^s) \tag{4.5}$$

$$\dot{x}_3 = -\eta_1 x_3 + \eta_2\left[\cos(x_1 + \delta^s) - \cos\delta^s\right] \tag{4.6}$$

where $K_1 = \dfrac{E}{(x_{12} + x'_d)}$, $\eta_1 = \dfrac{x_{12} + x_d}{(x_{12} + x'_d)T'_{do}}$

$$\eta_2 = \frac{(x_d - x'_d)\,E}{(x_{12} + x'_d)\,T'_{do}}$$

The energy function can be developed by using a modification of the Moore-Anderson theorem [Pai and Rai (1974)] or in any other manner such as the operational transfer technique [Rau and Krishnamoorthy (1983)] or the state function method [Miyagi and Taniguchi (1977)]. It is given by

$$\begin{aligned} V = {} & \frac{1}{2} M x_2^{\,2} - K_1(x_3 + e)\cos(x_1 + \delta^s) \\ & - P_i x_1 + K_1(x_3 + e)\cos\delta^s + \frac{1}{2}\frac{\eta_1}{\eta_2} K_1 x_3^{\,2} \end{aligned} \tag{4.7}$$

with $\dot{V} = \dfrac{-K_1}{\eta_2}(\dot{x}_3)^2$.

Siddiqee [1968] originally conjectured, by an analogy with a single machine system, that the closest u.e.p. is given by

$$x^u = \begin{bmatrix} \Pi - 2\,\delta^s \\ 0 \\ 0 \end{bmatrix}$$

The region of attraction of $x_1 = x_2 = x_3 = 0$ was therefore approximated as $V < V_{cr}^{(1)}$ where

$$V_{cr}^{(1)} = V\Big|_{x^u} = 2K_1\ e \cos \delta^s - P_o\ (\Pi - 2\,\delta^s) \qquad (4.8)$$

It was shown by Willems [1969] that this region was not closed and there are other u.e.p.'s closer to the origin than the one given above. The u.e.p.'s are the solutions of the algebraic equations obtained from (4.4)-(4.6) by setting the derivatives equal to zero.

$$\begin{aligned} x_2 &= 0 \\ (x_3 + e) \sin (x_1 + \delta^s) &= e \sin \delta^s \\ \eta_1\, x_3 + \eta_2 \left[\cos \delta^s - \cos (x_1 + \delta^s)\right] &= 0 \end{aligned} \qquad (4.9)$$

Besides x^u which satisfies (4.9) there is another closer u.e.p. given by

$$\hat{x}^u = \begin{bmatrix} \hat{x}_1 \\ 0 \\ \hat{x}_3 \end{bmatrix}$$

where $\hat{x}_1$ is the solution of

$$\sin (x_1 + \delta^s)\ \frac{\left[\cos \delta^s - \cos (x_1 + \delta^s)\right]}{\left[\sin (x_1 + \delta^s) - \sin \delta^s\right]} = \frac{e\,\eta_1}{\eta_2} \qquad (4.10)$$

with $\hat{x}_2 = 0$ and $\hat{x}_3 = \dfrac{\eta_2}{\eta_1}\left[\cos (x_1 + \delta^s) - \cos \delta^s\right]$

The stability region is given by $V < V_{cr}^{(2)}$ where $V_{cr}^{(2)}$ is obtained by evaluating V at $\hat{x}^u$. Thus

$$V_{cr}^{(2)} = -K_1 \frac{\eta_1}{\eta_2} \hat{x}_3 (e + \frac{\hat{x}_3}{2}) - K_1 e \hat{x}_1 \sin \delta^s \tag{4.11}$$

Numerical Example

A single machine is connected to an infinite bus through two parallel transmission lines (Fig. 4.1) with H = 2.77 sec., $x_d' = 0.3$ p.u., $x_d = 1.15$ p.u., E = 1 p.u., $T_{do}' = 6.6$ sec., $E_{fd} = 1.22$ p.u., $P_i = 0.8$. The reactance of each transmission line is 0.2 p.u. A fault on one of the transmission lines is cleared by opening the circuit breaker at both ends of the transmission line. Find the critical clearing time t_{cr}.

The equations of the post-fault system are

$$\begin{aligned} \dot{x}_1 &= x_2 \\ \dot{x}_2 &= 68 \left[0.8 - 2(x_3 + 1.02) \sin (x_1 + 0.42)\right] \\ \dot{x}_3 &= 0.15 \left[-1.53 - 2.7\, x_3 + 1.7 \cos (x_1 + 0.42)\right] \end{aligned} \tag{4.12}$$

The post-fault steady state values are

$$\delta^s = 0.42 \text{ rad}, \ \dot{\delta} = 0, \ e = 1.02 \text{ p.u.}$$

The two nearest u.e.p's are

$$x^u = (2.3,\ 0,\ 0)^T \quad \text{and} \quad \hat{x}^u = (1.2,\ 0,\ -0.6)^T$$

$$V_{cr}^{(1)} = V\big|_{x=x^u} = 1.87 \quad \text{and} \quad V_{cr}^{(2)} = V\big|_{x=\hat{x}^u} = 0.39$$

The region $V < V_{cr}^{(1)}$ is unbounded and the region $V < V_{cr}^{(2)}$ is bounded and shown by hatched lines (Fig. 4.2). The critical clearing time t_{cr} computed by integrating the faulted equations until $V = V_{cr}^{(2)} = 0.39$ is approximately 0.22 sec as compared to 0.55 sec if we use $V_{cr}^{(1)} = 1.87$.

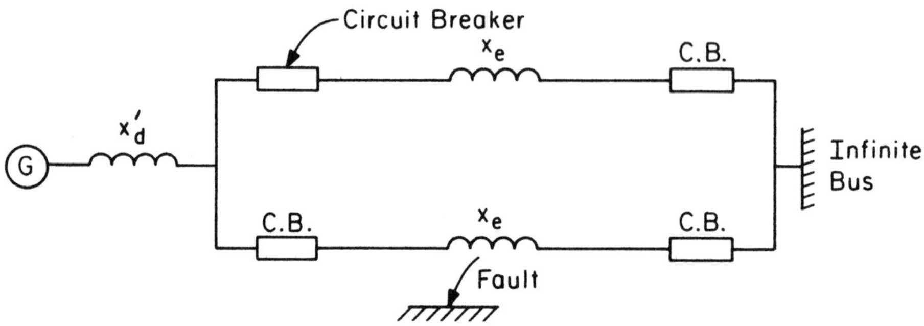

Figure 4.1 Synchronous machine connected to an infinite bus.

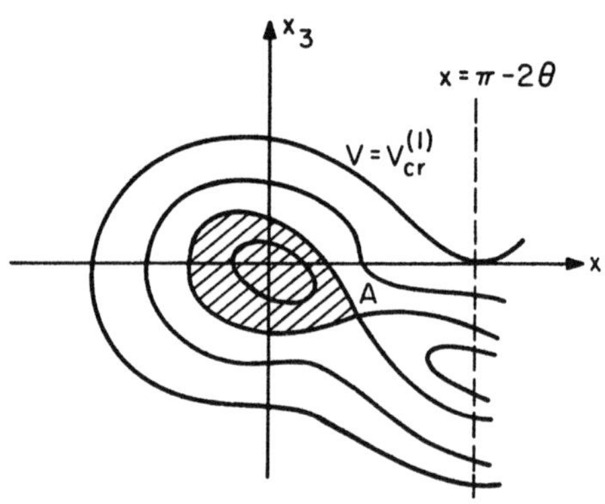

Figure 4.2 Region of stability of single machine system with flux decay model.

4.3 Multi-machine Systems with Flux Decay Model (Method of Parameter Variations)

This method of using parameter variation [Sasaki (1979)] to incorporate flux decay effects in a multi-machine stability investigation is a novel one and has been the basis of subsequent investigations for more detailed models by Athay et al. [1979]. In general it is difficult to construct Lyapunov functions analytically for more detailed models. Some alternatives are

1) Make approximation in the detailed model so that the resulting model can have an analytical energy/Lyapunov function.

2) Develop a numerical type energy function for the detailed model.

Sasaki's work in some sense falls in both categories as we explain below. He argues that if the traditional approach is followed, the value of E'_q in the u.e.p. falls below 0.4 which is physically not valid and also it is numerically difficult to find the u.e.p. Since in classical models E'_q is held constant at the pre-fault value, the u.e.p. obtained may not be the one of interest in multi-machine systems with flux decay. Hence, he proposed the method of parameter variations which treats E'_q as a parameter changing the voltage behind transient reactance of the classical model.

4.3.1 **Multi-machine Model**

If we include flux decay for each machine, there will be 3n first order differential equations. However, to simplify matters we take flux decay into account for the first p machines near the fault and only the classical model for the remaining n-p machines. The equations are

$$M_i \frac{d^2\delta'_i}{dt^2} + D_i \frac{d\delta'_i}{dt} = P_{mi} - \{E'_{qi} + (x_{qi} - x'_{di}) I_{di}\} I_{qi} \tag{4.13}$$

$$T'_{doi} \frac{dE'_{qi}}{dt} = E_{fd,i} - E'_{qi} - (x_{di} - x'_{di}) I_{di} \tag{4.14}$$

$$(i=1, 2, \ldots\ldots p)$$

$$M_i \frac{d^2\delta'_i}{dt^2} + D_i \frac{d\delta'_i}{dt} = P_{mi} - P_{ei} \qquad (i=p+1, \ldots\ldots n) \tag{4.15}$$

where $$P_{ei} = \sum_{j=1}^{n} |Y'_{ij}||E_i| \, [E'_{qj} + (x_{qj} - x_{dj})I_{dj}] \cos(\delta_{ij} - \phi_{ij})$$

$$+ \sum_{\ell=1}^{n} |Y'_{i\ell}||E_i||E_\ell| \cos(\delta_{i\ell} - \phi_{i\ell})$$

where $Y'_{ij} = |Y'_{ij}| \angle \phi_{ij}$ is the element of the internal node matrix computed on the basis of having voltage behind x_{qi} for i=1, ... p and x'_{di} for i = p+1, ... n. For details of the derivation of P_{ei} see [Sasaki (1979)].

Define $$E'^T_q = (E'_{q1}, E'_{q2} \, \ldots\ldots \, E'_{qp}) \text{ and } E^T_e = (E'^T_q, |E^T_{n-p}|)$$

where the components of $|E_{n-p}|$ are the magnitudes of the voltages behind transient reactances of the generators p+1, ... n. See the complex number representation of E'_{qi}, δ'_i, δ_i, etc. (Fig. 4.3).

We also consider the associated classical model (2.14) for the post-fault system shown below.

$$M_i \frac{d^2\delta_i}{dt^2} + D_i \frac{d\delta_i}{dt} = P_i - \sum_{\substack{j=1 \\ \neq i}}^{n} (C_{ij} \sin \delta_{ij} + D_{ij} \cos \delta_{ij}) \qquad i=1, 2 \ldots n \tag{4.16}$$

whose energy function is given by (2.19). Recall that $C_{ij} = |E_i||E_j| \, B_{ii}$ and $D_{ij} = |E_i||E_j| \, G_{ij}$. We now let $|E_i|$ in (4.16) be variable in magnitude for i=1, 2, ... p and the manner in which it is calculated is explained later. The energy function of the type (2.19) or (2.31) in the COI formulation can be used. Sasaki's method does not create an energy function for the system of Equations (4.13) - (4.15) but circumvents this problem by translating under some assumptions the changes in E'_{qi} in (4.14) into changes in $|E_i|$ (i=1, 2, ... p) in

(4.16). $|E_i|$ (i = p+1, ... n) are the magnitudes for the voltages behind transient reactances for the classical models of the (n-p) machines. For the first p machines, the parameter variation in $|E_i|$ in (4.16) is computed as explained below and is then correspondingly reflected into the energy function, critical value, etc.

4.3.2 Parametric approach

Consider the complex number representation (or the so-called phasor diagram) of the terminal voltage, currents and the voltage behind transient reactance of a machine (Fig. 4.3).

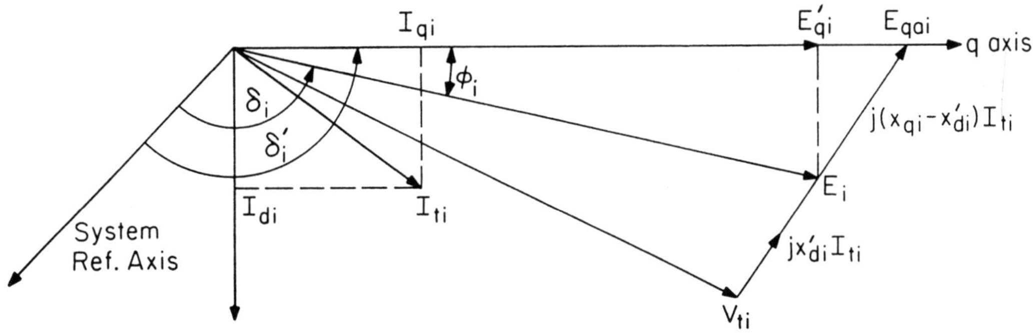

Figure 4.3 The phasor diagram of machine voltages and currents.

Assumptions

1) For the flux decay model, the angle ϕ_i between E'_{qi} and E_i before, during and after the fault remains constant, i.e.,

$$\text{Arg}\left(\frac{E'_{qi}}{E_i}\right) = \phi_i \quad \text{for each machine.}$$

2) $|E_i| = (k_i)\,E'_{qi}$ where $k_i = 1/\cos\phi_i$. Under these two assumptions the first p elements in $E = (|E_1|, \ldots\ldots |E_p|, |E_{p+1}|, \ldots\ldots |E_n|)^T$ are

$$|E_p| = k\,E'_q \quad , \quad k = \text{Diag}\,(k_i) \tag{4.17}$$

Define n×1 vector ϕ and n×n diagonal matrix T as

$$\phi^T = (\phi_1 \ldots\ldots \phi_p, 0 \ldots\ldots 0)$$

$$T = \text{Diag}(k_1 \ldots\ldots k_p, 1, 1, \ldots\ldots 1)$$

Define $\delta_i = \delta'_i - \phi_i$. Since ϕ_i is assumed constant for each machine

$$\frac{d\delta'_i}{dt} = \frac{d\delta_i}{dt}$$

Hence, we postulate the system equations in the classical form (4.16) with $|E_i|$ being varying for i=1, 2 .... p and determined from the faulted equations of the type (4.13) - (4.15) subject to the assumptions (1) and (2). In the energy function formed from (4.16), $|E_i|$ and δ_i are used in the usual manner. Thus, these equations in the "equivalent" classical form for the post-fault parameters are

$$M_i \frac{d^2\delta_i}{dt^2} + D_i \frac{d\delta_i}{dt} = P_{mi} - \sum_{j=1}^{n} |Y_{ij}||E_i||E_j| \cos(\delta_i - \delta_j - \delta_i^s + \delta_j^s - \phi_{ij}) \tag{4.18}$$

$$|E| = \begin{bmatrix} |E_1| \\ \cdot \\ \cdot \\ \cdot \\ |E_p| \\ \cdot \\ \cdot \\ |E_n| \end{bmatrix} = T\,|E_e| = T \begin{bmatrix} E'_q \\ |E_{n-p}| \end{bmatrix} \tag{4.19}$$

$$\delta = \delta' - \alpha \tag{4.20}$$

E'_q is a p-dimensional vector and is obtained from the numerical solution of (4.13) - (4.15). Since the post-fault equilibrium point δ^s and the unstable equilibrium point δ^u vary depending on $|E_i|$, these have be to recalculated at each time step. This is a major drawback of this method.

Having stated this parametrization procedure, it is relevant to reexamine the assumptions (1) and (2). For a fault close to the generator, the "phasor" diagram representation of Fig. 4.3 does not hold during fault for that generator. For the faulted generator, $V_{ti} = 0$ during the fault. Under fault condition, the generator under question is isolated from the remaining part of the system and has zero electrical output during the fault. Hence, acceleration for each machine is calculated solely on the basis of inertia. For machines far away from the fault, Figure (4.3) is more or less the same for the pre-fault and the post-fault systems.

The algorithm can be stated as follows:

1. Perform the pre-fault load flow calculation.
2. Determine Y matrix for the faulted system based on the internal node representation as in Fig. (4.4).

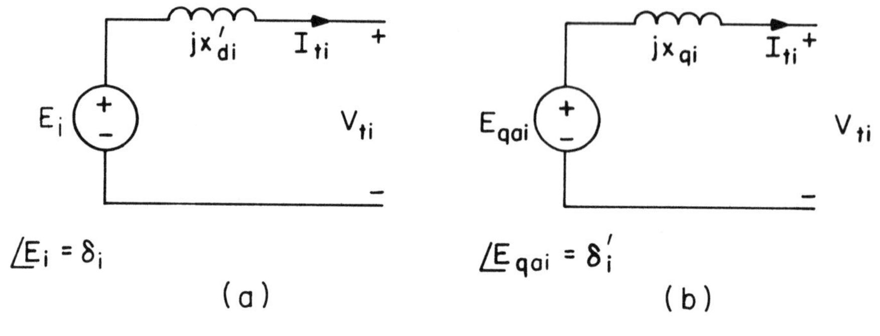

Figure 4.4. Internal node representation for the classical and flux decay models.

For the n-p generators, the internal node representation of Fig. (4.4a) is used and for the flux decay models Fig. (4.4b) is used. The internal nodal admittance matrix is calculated on this basis for the faulted system. The elements of the Y matrix are $Y'_{ij} = |Y'_{ij}| \angle\phi_{ij}$. Also calculate the Y matrix based on the nodes behind transient reactance for all the machines. This is used for the post-fault system.

3. Calculate the initial value E'_{qi} and angle ϕ_i of the p generators.

4. Obtain the stable equilibrium point (s.e.p) δ^s and the unstable equilibrium point (u.e.p) δ^u giving V_{cr} for the post-fault system (4.18) as $V_{cr} = V(\delta^s, \delta^u, |E|)$.

5. Execute forward numerical integration of the faulted system corresponding to (4.13) - (4.15). At each time step the vector E is computed from (4.19). Then recalculate $\delta^s = \delta^s(E)$ and δ^u (E) from (4.18) to update V_{cr}. The cricital clearing time, t_{cr}, is reached when $V = V_{cr}$ along the faulted trajectory. Clearly this step is very time consuming computationally.

4.3.3 Numerical Example [Sasaki (1979)]

The effectiveness of this approach was demonstrated on the 4-machine example (Fig. 4.5). Two fault locations are considered. G1 and G2 are treated as classical models (constant field flux linkage), whereas G3 and/or G4 are considered to be flux decay models. The line and load flow data are given in the paper by El-Abiad and Nagappan (1966). It is shown in Sasaki's paper that the parametric method of handling flux decay agrees well with the exact flux decay model by comparing the rotor angles in the post-fault period for a given fault.

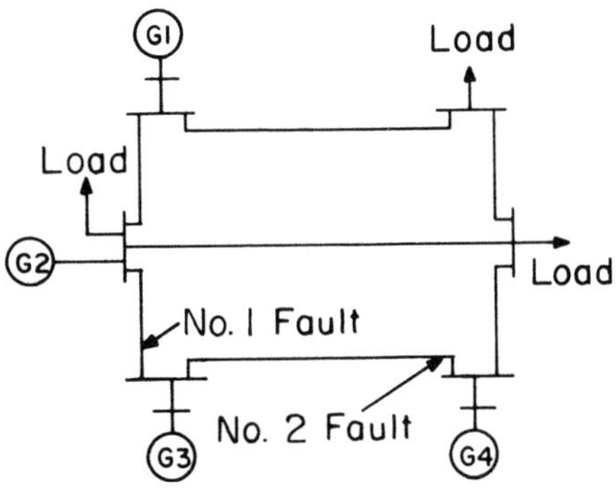

Figure 4.5 One-line diagram of the sample system

TABLE 4.1 SYNCHRONOUS MACHINE CONSTANTS

Generator	Capacity MVA	H Sec	D pu	X'_d pu	X_d pu	X_q pu	T'_{do} Sec
G1	100	10	1.0	0.04			
G2	15	1.5	6.0	1.0			
G3	40	3.0	2.5	0.5	1.45	0.95	8.0
G4	30	2.0	6.0	0.4	1.15	0.75	6.0

Tables 4.2 and 4.3 give the results with respect to computing t_{cr} for two fault cases. In each case two types of modeling are considered and the results shown.

(i) Column 3: G1 G2 G4 Classical model

G3 Flux decay model

(ii) Column 4: G1 G2 Classical model

G3 G4 Flux decay model

Note that when flux decay is considered both the s.e.p. and u.e.p. change. The results indicate that the approximations result in good agreement between t_{cr} by the direct method and the step-by-step simulation.

TABLE 4.2

Comparison between the Conventional Analysis (2nd column) and the Approximate Approach (3rd and 4th columns) for the Fault Case No. 1

Modeling	Constant Field Flux Linkage	G3: Field Flux Decay	G3, G4: Field Flux Decay
t_{act} (secs)	0.38	0.37	0.37
δ^s (rad)	(-0.109, -0.038, 0.321, 0.092)	(-0.119,-0.047, 0.360,0.093)	(-0.123,-0.051 0.368,0.103)
δ^u (rad)	(-0.614,-0.522, 2.528,-0.329)	(-0.608,-0.517, 2.497,-0.316)	(-0.614,-0.521, 2.504,-0.296)
V_c	1.010	0.867	0.843
t_{cr} (secs)	0.36	0.35	0.35

t_{act}: Reference critical fault clearing time by numerical integration

t_{cr}: Critical fault clearing time by the second method of Lyapunov

$\delta = (\delta_1, \delta_2, \delta_3, \delta_4)$

TABLE 4.3

Comparison between the Conventional Analysis (2nd column) and the Approximate Approach (3rd and 4th columns) for the Fault Case No. 2

Modeling	Constant Field Flux Linkage	G3, G4: Field Flux Decay	G4: Field Flux Decay
t_{act} (secs)	0.48	0.47	0.48
δ^s (rad)	(-0.054, 0.032, 0.173, -0.014)	(-0.063, 0.026, 0.175, 0.034)	(-0.061, 0.027 0.166, 0.037)
δ^u (rad)	(-0.580,-0.376, 2.458,-0.506)	(-0.589,-0.383, 2.450,-0.443)	(-0.592,-0.384, 2.459,-0.442)
V_c	1.508	1.432	1.499
t_{cr} (secs)	0.40	0.42	0.43

4.4 Lyapunov Functions for Multi-machine Systems with Flux Decay Model

In this section we essentially use the same approximations but derive rigorously a Lyapunov function for the approximate multi-machine flux decay model after neglecting the transfer conductances. The method is a generalization of the method due to Pai and Rai [1974] for single machine systems and uses the matrix version of the Moore-Anderson theorem modified by the Desoer-Wu condition [1969]. The treatment closely follows that by Kakimoto, et al. [1979].

4.4.1 Stability criterion

Consider the state space model

$$\dot{x} = A\,x - B\,F(\sigma) \tag{4.21}$$

$$\sigma = C^T x$$

$x \varepsilon R^n$ $\sigma \varepsilon R^m$, A, B are matrices of appropriate order. The nonlinearity $F(\sigma)$ satisfies

1) $F(\sigma)$ is continuous and maps R^m into R^m.

2) $F(\sigma)$ also satisfies $F(\sigma)^T N \sigma \geq 0$ for some real constant matrix N for all $\sigma \in R^m$ and $F(\sigma) = 0$ if $\sigma = 0$.

3) There is a function $V_1 \in C'$ mapping R^m into R such that

$$V_1(\sigma) \geq 0 \qquad \text{for all } \sigma \in R^m \tag{4.22}$$

and $$V_1(\sigma) = 0 \qquad \text{if } \sigma = 0$$

and for some constant real matrix Q

$$\nabla V_1(\sigma) = Q^T F(\sigma) \qquad \text{for all } \sigma \in R^m. \tag{4.23}$$

The transfer function matrix of the system is given by

$$\frac{\sigma(s)}{\mathcal{L}[F(\sigma)]} \triangleq W(s) = C^T (sI - A)^{-1} B \tag{4.24}$$

where stands for the Laplace transform operation with $W(\infty) = 0$. The figure (4.6) shows the transfer function matrix.

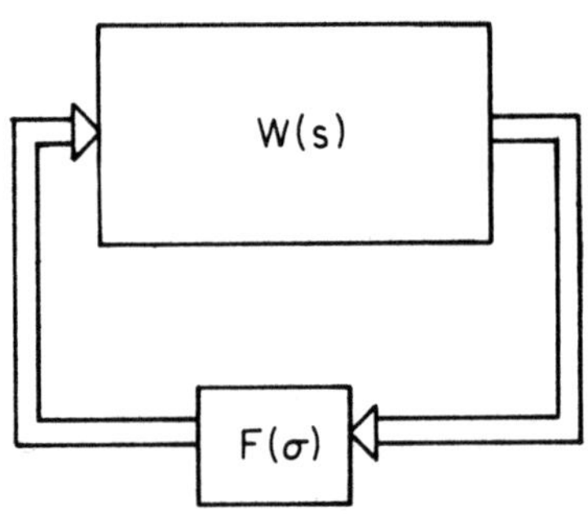

Figure 4.6 Block diagram of the nonlinear system (4.21)

Theorem [Anderson (1967)]

If there exists real matrices N and Q such that (N + Qs) W(s) is positive real, then the system shown in Fig. (4.6) is stable provided (N + Qs) does not cause pole-zero cancellation in W(s). Then a Lyapunov function for the system exists of the form

$$V(x) = x^T P x + 2V_1(\sigma). \tag{4.25}$$

4.4.2 Power system model [Sauer (1987)]

Using the 2 axis model and referring to Fig. 4.7

$$T'_{qoi} \frac{dE'_{di}}{dt} = - E'_{di} + (x_{qi} - x'_{qi}) I_{qi} \tag{4.26}$$

$$T'_{doi} \frac{dE'_{qi}}{dt} = - E'_{qi} - (x_{di} - x'_{di}) I_{di} + E_{fdi} \tag{4.27}$$

$$\frac{2H_i}{\omega_o} \frac{d\omega_i}{dt} = P_{mi} - \left[E'_{di} I_{di} + E'_{qi} I_{qi}\right] + (x'_{qi} - x'_{di}) I_{di} I_{qi} - D_i (\omega_i - \omega_o) \tag{4.28}$$

$$\frac{d\delta'_i}{dt} = \omega_i - \omega_o \qquad i=1, 2, \ldots n. \tag{4.29}$$

The electric power output of the ith machine can also be identified in terms of the equivalent circuit (Fig. 4.7).

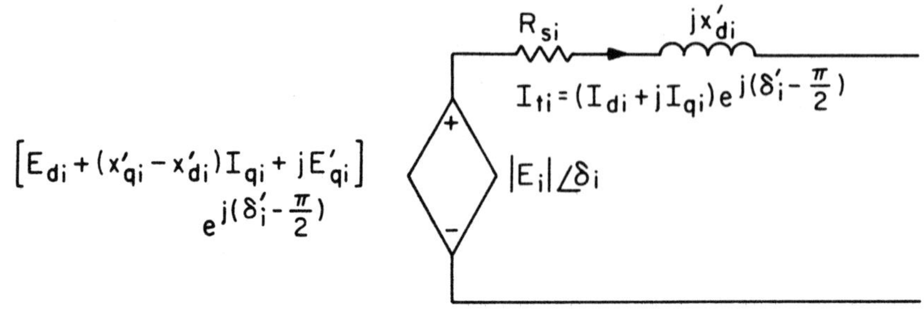

Figure 4.7 Equivalent circuit of ith machine.

If $R_{si} = 0$, then for n machines with the reduced internal Y matrix we can derive (as in Chapter 2)

$$P_{gi} = \sum_{j=1}^{n} Y_{ij} |E_i||E_j| \cos(\delta_i - \delta_j - \phi_{ij}) \tag{4.30}$$

where $\quad Y_{ij} = |Y_{ij}| \angle \phi_{ij}$.

Hence, the equations become (with $M_i = \frac{2H_i}{\omega_o}$)

$$\begin{aligned} M_i \frac{d\omega_i}{dt} &= P_{mi} - \sum_{j=1}^{n} |Y_{ij}||E_i||E_j| \cos(\delta_i - \delta_j - \phi_{ij}) \\ &\qquad - D_i (\omega_i - \omega_o) \\ &= P_{mi} - \sum_{j=1}^{n} |Y_{ij}||E_i||E_j| \sin(\delta_i - \delta_j + \theta_{ij}) \\ &\qquad - D_i (\omega_i - \omega_o) \end{aligned} \tag{4.31}$$

$$\text{with } \theta_{ij} = \frac{\pi}{2} - \phi_{ij} \qquad i=1, 2, \ldots n.$$

Let $T'_{qoi} \to 0$ so that we can eliminate the damper winding

$$E'_{di} = (x_{qi} - x'_{qi}) I_{qi} \tag{4.32}$$

$$T'_{doi} \frac{dE'_{qi}}{dt} = E_{fdi} - E'_{qi} - (x_{di} - x'_{di}) I_{di} \tag{4.33}$$

$$M_i \frac{d^2\delta'_i}{dt^2} + D_i \frac{d\delta'_i}{dt} = P_{mi} - \sum_{j=1}^{n} |Y_{ij}||E_i||E_j| \sin(\delta_i - \delta_j + \theta_{ij}) \qquad i=1, 2, \ldots n \tag{4.34}$$

where I_{di}, I_{qi}, δ_i, $|E_i|$ are found from the network equations and the equivalent circuit of Fig. 4.7 is modified as in Fig. 4.8.

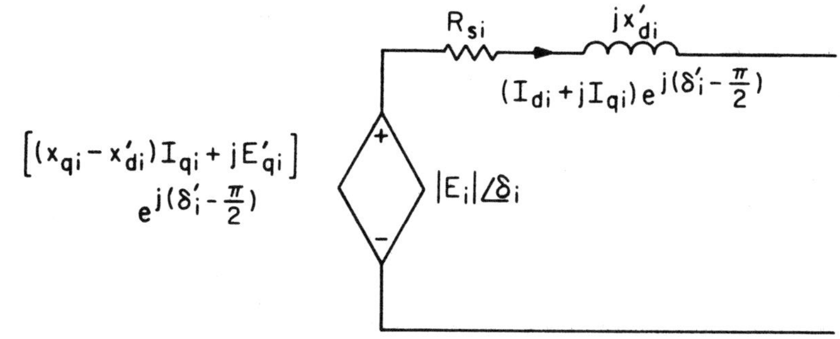

Figure 4.8 Modified equivalent circuit.

Clearly $|E_i|\angle\delta_i$

$$= \sqrt{(x_{qi}-x'_{di})^2 I_{qi}^2 + E'^2_{qi}} \; \angle \delta'_i - \frac{\pi}{2} + \tan^{-1}\frac{E'_{qi}}{(x_{qi}-x'_{di})\, I_{qi}} \tag{4.35}$$

The assumption $\delta'_i-\delta_i = \phi_i$ = constant requires $\frac{\pi}{2} - \tan^{-1}\frac{E'_{qi}}{(x_{qi}-x'_{di})\, I_{qi}}$ = constant

which means $(x_{qi} - x'_{di})\, I_{qi} = k_i\, E'_{qi}$ where k_i is a constant. With this

$$|E_i| = \sqrt{k_i^2 + 1}\,(E'_{qi}), \quad \phi_i = \frac{\pi}{2} - \tan^{-1}\frac{1}{k_i} = \frac{\pi}{2} - \alpha_i \tag{4.36}$$

Since $\tan\alpha_i = \frac{1}{k_i}$ it is easy to verify $\frac{1}{\sqrt{1+k_i^2}} = \sin\alpha_i = \cos\phi_i$. Therefore

$$E'_{qi} = |E_i| \cos\phi_i \tag{4.37}$$

and $\qquad \delta_i = \delta'_i - \phi_i$

Now solve for I_{di} in terms of E_i. From I = YE where Y is the reduced nodal admittance matrix at the internal nodes we get

$$(I_{di} + j\, I_{qi})\, e^{j(\delta_i' - \pi/2)} = \sum_{j=1}^{n} |Y_{ij}| \angle \phi_{ij}\, |E_j| \angle \delta_j \tag{4.38}$$

$$I_{di} + j\, I_{qi} = e^{-j(\delta_i' - \pi/2)} \sum_{j=1}^{n} |Y_{ij}| \angle \phi_{ij}\, |E_j| \angle \delta_j \tag{4.39}$$

Equating the real parts (algebra omitted) and noting $\delta_i' = \delta_i + \phi_i$, $\phi_{ij} = \frac{\pi}{2} - \theta_{ij}$, we get

$$I_{di} = -\sum_{j=1}^{n} [\, |Y_{ij}|\,|E_j| \cos(\delta_{ij} + \theta_{ij}) \cos \phi_i - |Y_{ij}|\,|E_j| \sin(\delta_{ij} + \theta_{ij}) \sin \phi_i] \tag{4.40}$$

Substituting (4.37) and (4.40) in (4.33) and noting that ϕ_i = constant

$$T'_{doi} \frac{d|E_i|}{dt} = [\, \frac{E_{fdi}}{\cos\phi_i} - |E_i| + (x_{di} - x'_{di}) \sum_{j=1}^{n} |Y_{ij}|\,|E_j| \cos(\delta_{ij} + \theta_{ij}) - \tan\phi_i\, (x_{di} - x'_{di}) \sum_{j=1}^{n} |Y_{ij}|\,|E_j \sin(\delta_{ij} + \theta_{ij})] \tag{4.41}$$

$$M_i \frac{d^2\delta_i}{dt^2} + D_i \frac{d\delta_i}{dt} = P_{mi} - \sum_{j=1}^{n} |Y_{ij}|\,|E_i|\,|E_j| \sin(\delta_{ij} + \theta_{ij}) \quad i=1, 2, \ldots n. \tag{4.42}$$

The rest of the treatment follows that in Kakimoto et al. (1979). We now make the following assumptions:

(i) For each model ϕ_i is small so that the term involving $\tan \phi_i$ can be neglected.

(ii) Neglect the transfer conductances, i.e., $|Y_{ij}| = B_{ij}$ and $\theta_{ij} = 0$. The equations (4.41) and (4.42) then become after a further assumption about $\cos \phi_i \approx 1$,

$$M_i \frac{d^2\delta_i}{dt^2} + D_i \frac{d\delta_i}{dt} = P_{mi} - \sum_{j=1}^{n} |E_i||E_j| B_{ij} \sin \delta_{ij}$$

$$T'_{doi} \frac{d|E_i|}{dt} = E_{fd,i} - |E_i| + (x_{di} - x'_{di}) \sum_{j=1}^{n} B_{ij} |E_j| \cos \delta_{ij}$$

$$i=1, 2, \ldots n. \qquad (4.43)$$

If δ_i^s, E_i^s are the equilibrium points of the above equations (written for the post-fault system) (4.43) can be written after transferring the equilibrium point to the origin as

$$M_i \frac{d^2\delta_i}{dt^2} + D_i \frac{d\delta_i}{dt} = \sum_{j=1}^{n} B_{ij} \left(|E_i^s||E_j^s| \sin \delta_{ij}{}^s - |E_i||E_j| \sin \delta_{ij}\right)$$

$$\frac{d|E_i|}{dt} = -\alpha_i \left(|E_i| - |E_i{}^s|\right) - \beta_i \sum_{\substack{j=1 \\ \neq 1}}^{n} B_{ij} \left(|E_j|^s \cos \delta_{ij}{}^s - |E_j| \cos \delta_{ij}\right) \qquad (4.44)$$

$$i=1, 2, \ldots n$$

$$\alpha_i = \frac{1 - (x_{di} - x'_{di}) B_{ii}}{T'_{doi}}, \quad \beta_i = \frac{x_{di} - x'_{di}}{T'_{doi}}$$

$$B_{ij} = |Y_{ij}| \cos \theta_{ij}$$

The above equations can be cast in the form of Eq. (4.21) with the state vector $x = (\delta_r^T, \omega^T, \Delta E^T)^T$ where the following (3n-1) variables are chosen as the state variables.

$$\delta_{ri} = \delta_{1(i+1)} - \delta^s_{1(i+1)} \qquad i=1, 2 \ldots n-1$$

$$\omega_i = \dot{\delta}_i \qquad i=1, 2 \ldots n \qquad (4.45)$$

$$\Delta E_i = E_i - E_i^s \qquad i=1, 2 \ldots n$$

δ_1 is taken as the reference angle for the angle subvector. Thus the state equations are in the minimal state space. The state equations of the system are given by

$$\dot{x} = Ax - BF(\sigma) \qquad (4.46)$$
$$\sigma = C^T x$$

where

$$A = \begin{bmatrix} 0 & K' & 0 \\ 0 & -M^{-1}D & 0 \\ 0 & 0 & -\alpha \end{bmatrix} \qquad B = \begin{bmatrix} 0 & 0 \\ M^{-1}T & 0 \\ 0 & \beta \end{bmatrix}$$

$$C = \begin{bmatrix} G & 0 \\ 0 & 0 \\ 0 & I \end{bmatrix}$$

and the various submatrices are

$$K' = \begin{matrix} 1 \\ n-1 \end{matrix} \begin{bmatrix} e \\ -I \end{bmatrix}_{n-1}$$

e is a row vector with all elements equal to 1.

$$T = \begin{matrix} \\ 1 \\ n-1 \end{matrix} \begin{matrix} \begin{matrix} n-1 & m-n+1 \end{matrix} \\ \begin{bmatrix} e & 0 \\ -I & T \end{bmatrix} \end{matrix} \qquad G = n-1 \overset{\begin{matrix} n-1 & m-n+1 \end{matrix}}{\begin{bmatrix} I & -T \end{bmatrix}}$$

T is related to K and G as

$$T = KG$$

and

$$m = \frac{n(n-1)}{2}$$

The nonlinearity consists of two vectors, i.e., $F(\sigma) = \begin{bmatrix} f_1(\sigma), \\ f_2(\sigma) \end{bmatrix}$. $f_1(\sigma)$ is an m-vector defined by

$$f_{1k}(\sigma) = B_{ij}\left(|E_i||E_j| \sin(\sigma_k + \delta_{ij}^s) - |E_i^s||E_j^s| \sin \delta_{ij}^s\right)$$

for i=1, 2 ... n-1, j=i+1, ... n and k=1, 2 ... m. (4.47)

k is related to i and j by

$$k = (i-1)n - \frac{i(i+1)}{2} + j \tag{4.48}$$

$f_2(\sigma)$ is an n-vector defined by

$$f_{2i}(\sigma) = \sum_{j=1}^{n} B_{ij}\left[|E_j|^s \cos \delta_{ij}^s - |E_j| \cos \delta_{ij}\right] \tag{4.49}$$

i=1, 2 ... n.

The output σ is an (m+n) vector defined by

$$\sigma_k = \delta_{ij} - \delta_{ij}^s \qquad k = 1, 2 \ldots m$$

$$\sigma = E_i - E_i^s \qquad k = m+1, \ldots m+n \tag{4.50}$$

k is related to i and j by (4.48) for k = 1, 2 ... m.

4.4.3 Lyapunov function

The state space model is now in the multivariable form with multiple memoryless coupled nonlinearities. As shown in Kakimoto et al. [1979] the multiplier (N+Qs) is chosen such that N = 0 and s = I so that

$$\nabla V_1(\sigma) = F(\sigma) \tag{4.51}$$

The function $V_1(\sigma)$ is chosen as follows.

$$V_1(\sigma) = \sum_{k=1}^{m} \int_0^{\sigma_k} f_{1k}(\sigma)\, d\sigma_k \tag{4.52}$$

$$= \sum_{i=1}^{n-1} \sum_{j=i+1}^{n} B_{ij}\big(|E_i||E_j| (\cos \delta_{ij}^s - \cos \delta_{ij})) - (\delta_{ij} - \delta_{ij}^s) |E_i^s||E_j^s| \sin \delta_{ij}^s \,. \tag{4.53}$$

The partial derivatives of $V_1(\sigma)$ are written as follows

$$\frac{\partial V_1}{\partial \sigma_k} = B_{ij} \left[|E_i||E_j| \sin(\sigma_k + \delta_{ij}^s) - |E_i^s||E_j^s| \sin \delta_{ij}^s \right]$$

$$= f_{1k}(\sigma) \quad \text{for } k = 1, 2 \ldots m \tag{4.54}$$

$$\frac{\partial V_1(\sigma)}{\partial \sigma_k} = \sum_{j=1}^{n} B_{ij} |E_j| (\cos \delta_{ij}^s - \cos \delta_{ij})$$

$$= f_{2k}(\sigma) \quad \text{for } k = m+1, \ldots m+n \tag{4.55}$$

Thus, (4.51) is satisfied with N = I, the identity matrix. Following Willem's method [1971], which involves considerable algebra, the P matrix in (4.25) is solved for and the Lyapunov function for the non-uniform damping case is obtained as (with $M_T = \sum_{i=1}^{n} M_i$)

$$V(x) = \left(\frac{1}{2M_T}\right) \sum_{i=1}^{n} \sum_{j=1}^{n} M_i M_j (\omega_i - \omega_j)^2$$

$$+ (\mu^* - \mu_o)\Big(\sum_{i=1}^{n} M_i \omega_i \Big)^2$$

$$+ \rho \Big\{ \sum_{i=1}^{n} \big(D_i(\delta_i - \delta_i^s) + \sum_{i=1}^{n} M_i \omega_i \big) \Big\}^2$$

$$+ \sum_{i=1}^{n} (\alpha_i/\beta_i)(|E_i| - |E_i^s|)^2 + \sum_{i=1}^{n} \sum_{j=1}^{n} B_{ij} \left[|E_i||E_j| (\cos \delta_{ij}^s - \cos \delta_{ij}) - (\delta_{ij} - \delta_{ij}^s) |E_i^s||E_j^s| \sin \delta_{ij}^s \right] \tag{4.56}$$

where ρ is a non-negative scalar. For simplicity sake, it is chosen to be zero. μ^* must lie between the two roots of the quadratic equation

$$(\mu^*)^2 \sum_{i=1}^{n-1} \sum_{j=i+1}^{n} \frac{(D_i M_j - D_j M_i)^2}{4 D_i D_j} - \mu^* \sum_{i=1}^{n} M_i - 1 = 0 \tag{4.57}$$

If damping is uniform $\mu^* = \mu_o = -1/\sum_{i=1}^{n} M_i$. The fourth term in (4.56) represents the effect of the flux decay while the fifth term is the potential energy term. If the damping is uniform or zero, the Lyapunov function becomes

$$\begin{aligned} V(x) &= \left(\frac{1}{2\sum_{i=1}^{n} M_i}\right)\left(\sum_{i=1}^{n} \sum_{j=1}^{n} M_i M_j (\omega_i - \omega_j)^2\right. \\ &+ \sum_{i=1}^{n} \sum_{j=1}^{n} B_{ij} \left[E_i E_j (\cos \delta_{ij}^s - \cos \delta_{ij})\right. \\ &\qquad \left. - (\delta_{ij} - \delta_{ij}^s) |E_i^s||E_j^s| \sin \delta_{ij}^s\right] \\ &+ \sum_{i=1}^{n} \left(\frac{\alpha_i}{\beta_i}\right) (|E_i| - |E_i^s|)^2 \\ &= V_k(\omega) + V_p(\delta, E) + V_f(E) \end{aligned} \tag{4.58}$$

If $\dot{V}(x)$ is evaluated along the system equations with damping present it can be verified that

$$\dot{V}(x) = -\left(\frac{1}{\sum_{i=1}^{n} D_i}\right) \sum_{i=1}^{n} \sum_{j=1}^{n} D_i D_j (\omega_i - \omega_j)^2 - 2 \sum_{i=1}^{n} \left(\frac{1}{\beta_i}\right)\left(\frac{d|E_i|}{dt}\right)^2 \tag{4.59}$$

Hence, V(x) is a valid Lyapunov function.

4.5 Multi Machine Systems with Flux Decay Effects and AVR [Miyagi and Bergen (1986)]

This work is an extension of the work of Kakimoto et al. (1980) to include the effects of AVR. The dynamic model of the ith machine is given by Eq. (4.43) as

$$M_i \frac{d^2\delta_i}{dt^2} + D_i \frac{d\delta_i}{dt} = P_{mi} - \sum_{j=1}^{n} B_{ij} |E_i||E_j| \sin \delta_{ij}$$

$$T'_{doi} \frac{d|E_i|}{dt} = E_{fdi} - |E_i| - (x_{di} - x'_{di}) I_{di} \tag{4.60}$$

We assume that the AVR is represented by a linear first order response

$$T_{vi} \frac{dE_{fdi}}{dt} = (E^s_{fdi} - E_{fdi}) - \mu_i (V_{ti} - V^s_{ti})$$

where T_{vi} = time constant of the voltage regulating system, μ_i is the feed back gain of the voltage regulating system, and V_{ti} is the magnitude of the terminal voltage.

Since $V_{ti} = \sqrt{V^2_{di} + V^2_{qi}}$, this introduces a difficulty in constructing the Lyapunov function. The following approximation is made, namely $V_{ti} = k_i V_{tqi}$ where k_i is a positive constant and V_{tqi} is the quadrature component. Thus using the relationship $V_{tqi} = - x'_{di} I_{di} + |E_i|$, we obtain V_{tqi} as a function of both E_i and δ_i. Shifting the post-fault equilibrium δ^s_i, $|E_i|^s$, E^s_{fdi}, $|E_i|^s$ to the origin, the dynamic model (4.60) is written as

$$M_i \frac{d^2\delta_i}{dt^2} + D_i \frac{d\delta_i}{dt} = \sum_{j=1}^{n} B_{ij} \{ |E_i|^s |E_j|^s \sin \delta^s_{ij} - |E_i||E_j| \sin \delta_{ij} \}$$

$$\frac{d|E_i|}{dt} = - \alpha_i (|E_i| - |E_i|^s) - \beta_i \sum_{\substack{j=1 \\ \neq i}}^{n} B_{ij} [|E_j|^s \cos \delta^s_{ij} - |E_j| \cos \delta_{ij}]$$
$$+ \gamma_i (E_{fdi} - E^s_{fdi}) \tag{4.61}$$

$$\frac{dE_{fdi}}{dt} = -\xi_i (E_{fdi} - E_{fdi}^s) - \eta_i (|E_i| - |E_i|^s)$$

$$+ \phi_i \sum_{\substack{j=1 \\ \neq i}}^{n} B_{ij} [|E_j|^s \cos \delta_{ij}^s - |E_j| \cos \delta_{ij}]$$

where α_i and β_i are defined as earlier in (4.44) and

$$\gamma_i = \frac{1}{T'_{doi}}, \quad \xi_i = \frac{1}{T_{vi}}, \quad \eta_i = \frac{(1 + B_{ii} x'_{di})}{T_{vi}} \mu_i k_i$$

$$\phi_i = \frac{\mu_i k_i x'_{di}}{T_{vi}}$$

Following the procedure similar to that in the previous section, a Lyapunov function for the above dynamic model can be obtained. For details see Miyagi and Bergen (1986).

4.6 Energy Functions with Detailed Models

The ultimate objective is to have energy functions with detailed models of the synchronous machine, excitation system, loads, etc. There have been three principal efforts in this area.

1. To extend Sasaki's method of Sec. 4.3 (Athay [1979]).

2. To obtain numerical Lyapunov functions (Sauer et al. [1988]).

3. To obtain the analytical or quasi-analytical form of the energy function (Tsolas et al. [1985], Padiyar and Sastry [1987]).

We shall discuss these in detail in the remaining sections.

4.6.1 Extension of Sasaki's method for detailed machine models

The first attempt at using energy functions for detailed models is due to Athay et al. [1979] using the parameter variation method discussed in Sec. 4.3. We briefly review the method of Sasaki here in a symbolic manner first.

Let E'_q be the vector of voltages corresponding to the field flux linkages and x the vector of generator angles and speeds. For the faulted system assume a detailed model

$$\dot{E}'_q = h\ (E'_q,\ x)$$

$$\dot{x} = g\ (x,\ E'_q) \tag{4.62}$$

For the post-fault system assume a classical model where the internal voltage vector is varying and is computed from a solution of the above equations.

$$\dot{x} = f\ (x,\ E) \tag{4.63}$$

We construct energy function for this model instead of the detailed model. The following observations are from Athay et al. (1979) where the above approach was tried on the detailed 2-axis model together with the excitation system.

The equations for the 2-axis model of the generator are (in machine reference frame)

Algebraic Equations

$$\begin{bmatrix} V_{qi} \\ V_{di} \end{bmatrix} = \begin{bmatrix} E'_{qi} \\ E'_{di} \end{bmatrix} + \begin{bmatrix} -R_a & -x_d \\ x'_q & -R_a \end{bmatrix} \begin{bmatrix} I_{qi} \\ I_{di} \end{bmatrix} \tag{4.64}$$

Differential Equations

$$T'_{do} \frac{d}{dt} E'_{qi} = - E'_{qi} - (x_{di} - x'_{di}) I_{di} + E_{fdi}$$

$$T'_{qo} \frac{d}{dt} E'_{di} = - E'_{di} + (x_{qi} - x'_{qi}) I_{qi} \quad (4.65)$$

The field voltage E_{fdi} is supplied by the excitation system whose differential equations depend on the type of excitation system used. We convert voltage and current variables to a synchronous reference frame. Letting $\theta_i(\theta_i = \delta_i - \pi/2)$ be the angle between the generator d axis and the synchronous reference frame, then

$$(I_{di} + j I_{qi}) e^{j\theta i} = I_i \quad (4.66)$$

where I_i is the current in the synchronous reference frame. Neglecting transient saliency, i.e., $x'_d = x'_q$, the algebraic equations can be expressed as

$$V_i = E'_i - (R_{ai} + j x'_{di}) I_i \quad (4.67)$$

where $E'_i = (E'_{di} + j E'_{qi}) e^{j(\delta_i - \pi/2)}$ and $V_i = e_i + jf_i$ is the generator terminal voltage. Neglecting stator resistance, we can express the ith generator injected current I_i as

$$I_i = \frac{1}{x'_{di}} \{(E'_{qi} \sin \delta_i - E'_{di} \cos \delta_i - f_i)$$

$$- j (E'_{qi} \cos \delta_i + E'_{di} \sin \delta_i - e_i)\} \quad (4.68)$$

and the real power of the ith generator as

$$P_{gi} = \frac{1}{x'_{di}} \{E'_{qi}(e_i \sin \delta_i - f_i \cos \delta_i) - E'_{di}(e_i \cos \delta_i + f_i \sin \delta_i)\} \tag{4.69}$$

From the detailed model for the fault-on state, P_{gi} is computed and used in the energy function for the classical model for the post-fault system using the center of angle notation.

$$M_i \dot{\tilde{\omega}}_i = P_{mi} - P_{gi} - \frac{M_i}{M_T} P_{COI} \overset{\Delta}{=} f_i \tag{4.70}$$

$$\dot{\theta}_i = \tilde{\omega}_i \qquad i=1,\ 2,\ \ldots\ n$$

The transient energy function is defined as the sum of first integrals and is expressed as

$$V = \frac{1}{2} \sum_{i=1}^{n} M_i \tilde{\omega}_i^2 - \sum_{i=1}^{n} \int_{\theta_i^s}^{\theta_i} f_i \, d\theta_i \tag{4.71}$$

$$= V_{K \cdot E} + V_{P \cdot E}$$

f_i is dependent not only on the angles but also on E'_{qi} and E'_{di} and these are obtained from a trajectory of the detailed model. Equation (4.71) is therefore to be computed numerically.

The above technique was tried on a 10 machine system [See Appendix [A]] with the following assumptions.

1) A two axis model was assumed and the effects of saturation and transient saliency were ignored, i.e., $x'_d = x'_q$

2) All machines except unit #10 were assumed to have IEEE Type 1 excitation system.

3) Mechanical damping was neglected.

4) All loads were modeled as constant impedances.

Using the trapezoidal rule the potential energy $V_{P \cdot E}$ is computed. f_i and θ_i are obtained from a transient stability program for the faulted system.

In the earlier work on classical models, the test for instability was conjectured to be when $f^T(\theta-\theta^s)$ changed sign from negative to positive. This was found to be true for the case of detailed model also. However, the PEBS method failed to accurately predict the critical value of the transient energy. In the case of the classical model representation, the value of V_{PE} at crossing of the PEBS was relatively unchanged for different $t_{c\ell} > t_{cr}$. This was not found to be true using detailed models with the internal voltage as the varying parameter. A modified form of the controlling u.e.p method was also tried but with limited success. However, these initial efforts at using detailed modeling in the framework of a numerical Lyapunov function provided a basis for further work in this area.

4.6.2 Numerical Lyapunov function

In the previous section it was mentioned that the PEBS method does not accurately predict the critical energy using a sustained fault trajectory. In this section we introduce a modification which accurately predicts critical energy and hence t_{cr} using the PEBS method.

Detailed 2-axis model (Constant load impedances) [Sauer et al. (1988)]

The equations are given by

$$\dot{\delta}_i = \omega_i - \omega_o \tag{4.72}$$

$$M_i\dot{\omega}_i = P_{mi} - (E'_{di}I_{di} + E'_{qi}I_{qi}) - D_i(\omega_i - \omega_o) \tag{4.73}$$

$$T'_{doi}\dot{E}'_{qi} = - E'_{qi} - (x_{di} - x'_{di})\, I_{di} + E_{fdi} \tag{4.74}$$

$$T'_{qoi}\dot{E}'_{di} = - E'_{di} + (x_{qi} - x'_{qi})\, I_{qi} \tag{4.75}$$

$$T_{Fi}\,\dot{R}_{fi} = -\,R_{fi} + \left(K_{Fi}/T_{Fi}\right)\,E_{fdi} \tag{4.76}$$

$$T_{Ei}\,\dot{E}_{fdi} = -\,\left(K_{Ei} + S_{Ei}\right)\,E_{fdi} + V_{ri} \tag{4.77}$$

$$T_{Ai}\,\dot{V}_{ri} = K_{Ai}R_{fi} - \left(\frac{K_{Ai}K_{Fi}}{T_{Fi}}\right)\,E_{fdi} - V_{ri} - K_{Ai}V_{ti} + K_{Ai}\,V_{refi} \tag{4.78}$$

$$i=1,\ 2,\ \ldots\ n$$

If we neglect transient saliency ($x'_{di} = x'_{qi}$) and use constant impedance loads for simplicity, the machine current components I_{di} and I_{qi} and the machine terminal voltage V_{ti} are given in terms of the system state variables by the interface equations (Y_{ij}, σ_{ij} are magnitude and angle variables of the admittance matrix reduced to internal nodes)

$$I_{di} = \sum_{j=1}^{n} Y_{ij}\,\left(E'_{dj}\,\cos(\delta_{ji} + \sigma_{ij}) - E'_{qj}\,\sin(\delta_{ji} + \sigma_{ij})\right) \tag{4.79}$$

$$I_{qi} = \sum_{j=1}^{n} Y_{ij}\,\left(E'_{qj}\,\cos(\delta_{ji} + \sigma_{ij}) + E'_{dj}\,\sin(\delta_{ji} + \sigma_{ij})\right) \tag{4.80}$$

$$V_{ti} = \sqrt{V_{di}^2 + V_{qi}^2}$$

$$V_{di} = x'_{qi}I_{qi} + E'_{di} \tag{4.81}$$

$$V_{qi} = -\,x'_{di}I_{di} + E'_{qi}$$

The saturation function in the exciter is given by

$$S_{Ei} = A_{sat,i}\;e^{(B_{sat,i}\;E_{fdi})}$$

Defining $x_i = [E'_{qi}\ E'_{di}\ R_{fi}\ E_{fdi}\ V_{ri}]^T$

$$y_i = [I_{di},\ I_{qi},\ V_{di},\ V_{qi},\ V_{ti}]^T$$

The n machine model therefore has the general form

$$\dot{\delta}_i = \omega_i - \omega_o \tag{4.82}$$

$$M_i\dot{\omega}_i = f_i(\delta,\ \omega,\ x,\ y) \qquad i=1,\ 2,\ \ldots\ n \tag{4.83}$$

$$\dot{x}_i = g_i(\delta,\ \omega,\ x,\ y) \tag{4.84}$$

$$0 = h_i(\delta,\ \omega,\ x,\ y) \tag{4.85}$$

In this particular model, the vector y can be solved for explicitly in terms of the states δ, ω, and x.

Generalized potential energy

The center of inertia formulation is used so that

$$\delta_{COI} \triangleq \frac{1}{M_T}\sum_{i=1}^{n} M_i\ \delta_i\ ,\ \omega_{COI} \triangleq \frac{1}{M_T}\sum_{i=1}^{n} M_i\ \omega_i$$

where $M_T = \sum_{i=1}^{n} M_i$.

The transient kinetic energy is given by the expression

$$V_{TKE} = \frac{1}{2}\sum_{i=1}^{n} M_i(\omega_i - \omega_{COI})^2 \tag{4.86}$$

Clearly if V_{TKE} is monitored after fault clearing at $t = t_{c\ell}$ it will provide an indication of synchronous stability. If the system is stable then $V_{TKE} \to 0$ as $t \to \infty$. V_{TKE} is converted into some other form of energy which we shall call as generalized potential energy. A generalized potential energy V_{GPE} is now defined such that when it is added to V_{TKE} and both evaluated along a cleared (final configuration) system trajectory the total is constant with respect to time. This requires

$$V_{TKE} + V_{GPE} = E \text{ (a constant) for } t \geq t_{c\ell}.$$

This condition requires

$$\dot{V}_{KE} + \dot{V}_{GPE} = 0 \qquad \text{for } t \geq t_{c\ell} \tag{4.87}$$

or from (4.83) and (4.86)

$$\dot{V}_{GPE} = - \dot{V}_{KE} = - \sum_{i=1}^{n} (\omega_i - \omega_{COI})\, f_i(\delta, \omega, x, y) \tag{4.88}$$

where the function f_i and all variables δ, ω, x, y must pertain to the post-fault system. We now proceed to develop a technique based on the sustained fault trajectory. Since V_{GPE} is defined to be such that it is to be computed from the final configuration, then the y variables in (4.88) must be computed accordingly. This calls for a modification in V_{GPE}. We give below the model for computing V_{GPE} both during and after the fault is cleared. Till $t = t_{c\ell}$, we use δ, ω, and x obtained from the fault-on trajectory in V_{GPE} but f_i pertaining to post-fault configuration. Hence, $V_{TKE} + V_{GPE} \neq$ constant up to $t = t_{c\ell}$ but only true for $t > t_{c\ell}$. The complete model is given by

For $t > 0$

$$\dot{\delta} = \omega - \omega_o \qquad \delta(o) = \delta_o \tag{4.89}$$

$$M\,\dot{\omega} = f(\delta, \omega, x, y) \qquad \omega(o) = \omega_o \tag{4.90}$$

$$\dot{x} = g(\delta, \omega, x, y) \qquad x(o) = x_o \tag{4.91}$$

$$y = \phi(\delta,\ \omega,\ x,\ p^{f}) \tag{4.92}$$

$$\hat{y} = \phi(\delta,\ \omega,\ x,\ p^{pf}) \tag{4.93}$$

$$\dot{V}_{GPE} = -\sum_{i=1}^{n} (\omega_i - \omega_o)\ f_i(\delta,\ \omega,\ x,\ \hat{y}),\ V_{GPE}(0) = 0 \tag{4.94}$$

The initial condition on V_{GPE} may be taken to be equal to zero. The superscripts f and pf stand for faulted and post-fault, respectively, and $\hat{y}$ stands for post-fault solution of network variables.

The above formulation produces good results when we use the classical model. This was also the experience of Athay et al. (1979). However, the results of V_{cr} and t_{cr} were not good when using detailed models. Sauer et al. (1988) point out that the main cause for this discrepancy is the existence of fast dynamics in the system equations. This is mainly due to the fast damper windings whose dynamics decay to their steady state value in a very short time. They propose the following procedure to properly account for the fast dynamics. In Equations (4.89) - (4.94) x is partitioned into $\hat{x}_1$ and x_2 where $\hat{x}_1$ is a subvector of fast variables. The dynamics for the fast variables are then written for the post-disturbance configuration.

$$\dot{\delta} = \omega - \omega_o \qquad \delta(o) = \delta_o \tag{4.95}$$

$$M\,\dot{\omega} = f(\delta,\ \omega,\ \hat{x}_1,\ x_2,\ y) \qquad \omega(o) = \omega_o \tag{4.96}$$

$$\dot{\hat{x}}_1 = g_1\ (\delta,\ \omega,\ \hat{x}_1,\ x_2,\ \hat{y}) \qquad \hat{x}_1(o) = x_1^{o} \tag{4.97}$$

$$\dot{x}_2 = g_2(\delta,\ \omega,\ \hat{x}_1,\ x_2,\ y) \qquad x_2(o) = x_2^{o} \tag{4.98}$$

$$y = h\ (\delta,\ \omega,\ \hat{x}_1\ x_2,\ p^f) \tag{4.99}$$

$$\hat{y} = h\ (\delta,\ \omega,\ \hat{x}_1,\ x_2,\ p^{pf}) \tag{4.100}$$

$$\dot{V}_{GPE} = -\sum_{i=1}^{n} (\omega_i - \omega_{COI})\ f_i(\delta,\ \omega,\ \hat{x}_1,\ x_2,\ \hat{y}),\ V_{GPE}(0) = 0 \tag{4.101}$$

Using the above equations the value of V_{GPE}^{max} along a sustained fault trajectory agrees very well with the first V_{GPE} peak obtained by using a critically cleared trajectory. Further refinements in this method include a methodology to detect the peak in V_{GPE} along a sustained fault trajectory. Omitting details, the PEBS is crossed along a sustained fault trajectory if

$$\sum_{i=1}^{n} f_i(\delta,\ \omega,\ x,\ \hat{y})\ [(\delta_i - \delta_i^s) - (\delta_{COI} - \delta_{COI}^s)] = 0\ . \tag{4.102}$$

When this condition is met, V_{GPE} experiences the peak first, which can be taken as V_{cr}.

Algorithm and Results

1. Begin with initial conditions from some prefault stable equilibrium point.
2. Integrate the faulted equations as formulated above to take into account the fast dynamics.
3. Compute V_{TKE} and V_{GPE} at each time step.
4. Also test the PEBS crossing condition given by Eq. (4.102).
5. Stop when the PEBS crossing is detected. The value of V_{GPE} at this point is V_{cr}.
6. Return to the time history of $V_{TKE} + V_{GPE}$ during fault and find when this sum is equal to V_{cr}. This gives t_{cr}.

Results

The above algorithm has been successfully implemented on the 39 bus 10 machine systems (see Appendix A for data). The results are presented in Table 4.4.

TABLE 4.4

RESULTS ON THE 10 MACHINE SYSTEM

Fault	t_{cr} (secs) (simulation)	t_{cr} (secs) (direct method with sustained fault)
Bus 17 (no line cleared)	.23 (10.6)	.23 (12.7)
Bus 4 (clear 4-14)	.26 (10.8)	.25 (10.5)
Bus 24 (clear 24-23)	.22 (8.1)	.20 (8.2)
Bus 37 (no line cleared)	.23 (6.8)	.21 (7.8)
Bus 29 (clear 28-29)	.05 (.32)	.045 (.25)

The numbers in brackets are the values of V_{GPE}^{max} for the critically cleared case in the simulation and for the sustained fault case. In the direct method V_{GPE}^{max} is taken as V_{cr}.

4.6.3 Quasi-analytical expressions for energy functions with detailed models

We consider the 2-axis model as in the previous section and try to construct a quasi-analytical energy function. Originally, the flux decay model in a structure preserving framework with no local loads at generator buses and constant P, Q loads at other buses was considered by Tsolas et al. (1985). It was generalized to include the excitation system by Padiyar and Sastry (1987). The extension to include a q axis damper winding and retaining transient saliency along with transmission line resistances is due to Padiyar and Ghosh (1989). From this the special cases can be easily derived. There is a difference in notations adopted in the above cited work as compared to the previous section. Whereas the previous section used the industry notation, these papers use the IEEE recommended notation which is not used by the industry. The conversion between these two notations is discussed below.

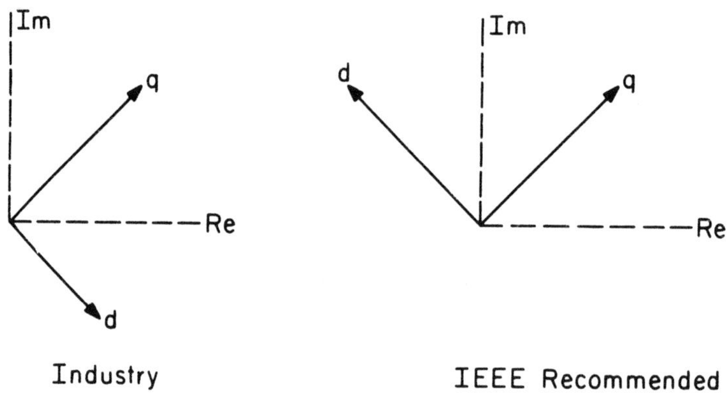

Figure 4.9 Two systems of d-q axis representation.

$$\begin{bmatrix} Re \\ Im \end{bmatrix} = \begin{bmatrix} \sin\delta & \cos\delta \\ -\cos\delta & \sin\delta \end{bmatrix} \begin{bmatrix} d \\ q \end{bmatrix} \qquad \begin{bmatrix} Re \\ Im \end{bmatrix} = \begin{bmatrix} \cos\delta & -\sin\delta \\ \sin\delta & \cos\delta \end{bmatrix} \begin{bmatrix} q_E \\ d_E \end{bmatrix}$$

INDUSTRY IEEE NOTATION (4.103)

We now discuss the method in this new (IEEE recommended) convention and largely follow Padiyar and Ghosh (1989). Note from (4.103) that I_d, I_q components in the two conventions are related by the simple relation that the d variables are negative of each other whereas the q variables remain the same.

Generator Model

Consider an 'n' machine system supplying 'N' non-linear voltage dependent loads. Usually in transient stability studies, the generators close to the fault are modeled in greater detail. The complexity in modeling the generators more realistically depends on the number of rotor circuits represented. The model considered here includes two circuits on the rotor; i) the field winding on the direct axis and ii) a damper winding on the quadrature axis. The excitation system is represented by the standard IEEE Type I model. It is assumed that the mechanical input of the generator is constant. The dynamics of the machines are then described by the following differential-algebraic equations (with respect to COI reference frame):

$$\dot{\theta}_1 = \tilde{\omega}_i \qquad (4.104)$$

$$i = 1, 2 \ldots n$$

$$M_i \dot{\tilde{\omega}}_i = P_{mi} - P_{ei} - \frac{M_i}{M_T} P_{COI} \qquad (4.105)$$

where

$$P_{ei} = \left[E'_{qi} + (x'_{di} - x'_{qi}) I_{di}\right] I_{qi} + E'_{di} I_{di} \qquad (4.106)$$

$$P_{COI} = \sum_{i=1}^{m} (P_{mi} - P_{ei}) \qquad (4.107)$$

$$T'_{doi} \dot{E}'_{qi} = E_{fdi} - E'_{qi} + (x_{di} - x'_{di}) I_{di} \qquad (4.108)$$

$$T'_{qoi} \dot{E}'_{di} = -\left[E'_{di} + (x_{qi} - x'_{qi}) I_{qi}\right] \qquad (4.109)$$

$$V_{qi} = E'_{qi} + x'_{di} I_{di} \qquad (4.110)$$

$$V_{di} = E'_{di} - x'_{qi} I_{qi} \qquad (4.111)$$

$$V_i e^{j\phi i} = (V_{qi} + jV_{di}) e^{j\theta i} \qquad (4.112)$$

Note that ϕ is the phase angle of the generator bus with respect to the COI. From Eq. (4.112) we get,

$$\begin{aligned} V_{qi} &= V_i \cos(\theta_i - \phi_i) \\ V_{di} &= -V_i \sin(\theta_i - \phi_i) \end{aligned} \tag{4.113}$$

From Eqs. (4.110) - (4.113), I_{di} and I_{qi} are obtained as

$$\begin{aligned} I_{di} &= \frac{[V_i \cos(\theta_i - \phi_i) - E'_{qi}]}{x'_{di}} \\ I_{qi} &= \frac{[E'_{di} + V_i \sin(\theta_i - \phi_i)]}{x'_{qi}} \end{aligned} \tag{4.114}$$

Substituting for I_{di} and I_{qi} from Eq. (4.114) in Eq. (4.106) and simplifying, we get,

$$P_{ei} = \frac{E'_{qi} V_i \sin(\theta_i - \phi_i)}{x'_{di}} + \frac{E'_{di} V_i \cos(\theta_i - \phi_i)}{x'_{qi}} + V_i^2 \frac{x'_{di} - x'_{qi}}{2x'_{di} x'_{qi}} \sin 2(\theta_i - \phi_i) \tag{4.115}$$

Eliminating I_{di} and I_{qi} in Eqs. (4.108) and (4.109) using Eq. (4.114) we get,

$$T'_{doi} \dot{E}'_{qi} = E_{fdi} - (x_{di}/x'_{di}) E'_{qi} + V_i \cos(\theta_i - \phi_i)(x_{di} - x'_{di})/x'_{di} \tag{4.116}$$

$$T'_{qoi} \dot{E}'_{di} = - [\frac{x_{qi}}{x'_{qi}} E'_{di} + \frac{V_i \sin(\theta_i - \phi_i)}{x'_{qi}} (x_{qi} - x'_{qi})] \tag{4.117}$$

Excitation System

The equations for the IEEE Type I excitation (Fig. 4.10) are written as

$$V_{ei} = V_{ref,i} - V_i - V_{si} \tag{4.118}$$

$$\dot{V}_{ri} = (K_{Ai} V_{ei} - V_{ri})/T_{Ai} \tag{4.119}$$

$$\dot{E}_{fdi} = [V_{R\ell i} - A_{ex} e^{B_{ex} E_{fd,i}} - K_{Ei} E_{fd,i}]/T_{Ei} \tag{4.120}$$

$$V_{si} = \frac{K_{si}}{T_{si}} E_{fd,i} - V_{s\ell,i} \tag{4.121}$$

$$\dot{V}_{s\ell,i} = \frac{1}{T_{si}} \left(\frac{K_{si}}{T_{si}} E_{fd,i} - V_{s\ell,i}\right) \tag{4.122}$$

The last two equations are obtained from the partial fraction expansion of $SK_s/1+ST_s$ as $[\frac{K_s}{T_s} - \frac{K_s}{T_s(1+ST_s)}]$ and designating $V_{s\ell,i}$ as the output of the integrator. Equation (4.118)-(4.123) are equivalent to (4.76)-(4.78) with $V_{s\ell,i} = R_{fi}$ and the other gains and time constants are easily identified.

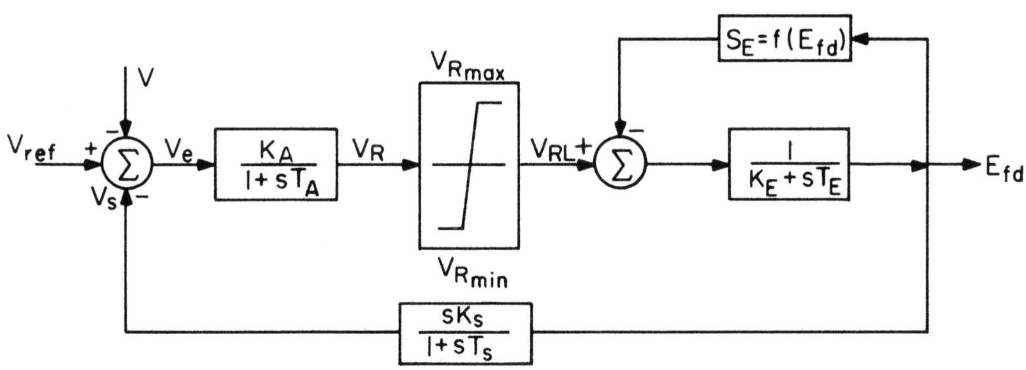

Figure 4.10 IEEE Type I excitation system.

Load Model

We use the same numbering as in Chapter 2; namely, the internal nodes 1, 2, ... n, generator buses n+1, ... 2n, load buses 2n+1, ... n+N. Load at each bus is represented by an arbitrary function of voltage at the respective bus. Thus,

$$P_{Li} = f_{pi}(V_i), \quad i = n+1, \ldots, n+N \tag{4.123}$$

$$Q_{Li} = f_{qi}(V_i), \quad i = n+1, \ldots, n+N \tag{4.124}$$

Power Flow Equations

For the lossless system the following equations can be written at bus i as the sum of real and reactive powers out of the branches where P_i = injected power into bus i.

$$P_i = \frac{E'_{qi} V_i \sin(\phi_i - \theta_i)}{x'_{di}} - \frac{E'_{di} V_i \cos(\theta_i - \phi_i)}{x'_{qi}}$$

$$+ \frac{V_i^2 (x'_{di} - x'_{qi})}{2x'_{di} x'_{qi}} \sin 2(\phi_i - \theta_i)$$

$$+ \sum_{j=n+1}^{n+N} B_{ij} V_i V_j \sin \phi_{ij}, \quad i = n+1, \ldots, 2n \tag{4.125}$$

$$= \sum_{j=n+1}^{n+N} B_{ij} V_i V_j \sin \phi_{ij}, \quad i = 2n+1, \ldots, n+N \tag{4.126}$$

$$Q_i = \frac{V_i^2}{x'_{di}} - \frac{E'_{qi} V_i \cos(\theta_i-\phi_i)}{x'_{di}} - \frac{E'_{di} V_i \sin(\phi_i-\theta_i)}{x'_{qi}}$$

$$- \frac{V_i^2(x'_{di} - x'_{qi})}{2x'_{di}x'_{qi}} [\cos 2(\theta_i-\phi_i) - 1]$$

$$- \sum_{j=n+1}^{n+N} B_{ij} V_i V_j \cos \phi_{ij}, \quad i = n+1, \ldots, 2n \tag{4.127}$$

$$= - \sum_{j=n+1}^{n+N} B_{ij} V_i V_j \cos \phi_{ij}, \quad i = 2n+1, \ldots, n+N \tag{4.128}$$

In the above expressions $B_{ij} = \text{Im}\,[Y_{ij}]$ where [Y] is the admittance matrix of the network (excluding machine reactances). Using the load model (4.123) and (4.124), P_i and Q_i in (4.125) - (4.128) can be written as

$$P_i = - f_{pi}(V_i), \quad i = n+1, \ldots, n+N \tag{4.129}$$

$$Q_i = - f_{qi}(V_i), \quad i = n+1, \ldots, n+N \tag{4.130}$$

<u>Structure Preserving Energy Function</u> (SPEF)

In what follows, a structure preserving energy function, which incorporates the effect of the excitation system, damper winding and voltage dependent active and reactive power loads, is developed.

Consider the function defined for the post-fault system,

$$W(\theta,\tilde{\omega},E'_q,E'_d,V,\phi,E_{fd},t) = W_1+W_2 = W_1 + \sum_{i=1}^{13} W_{2i} \tag{4.131}$$

where

$$W_1(\tilde{\omega}) = \frac{1}{2} \sum_{i=1}^{n} M_i\tilde{\omega}_i^2 \tag{4.132}$$

$$W_{21}(\theta) = -\sum_{i=1}^{n} P_{mi}(\theta_i - \theta_{io}) \tag{4.133}$$

$$W_{22}(t) = \sum_{i=n+1}^{n+N} \int_{t_o}^{t} f_{pi}(V_i) \frac{d\phi_i}{dt} dt \tag{4.134}$$

$$W_{23}(V) = \sum_{i=n+1}^{n+N} \int_{V_{io}}^{V_i} \frac{f_{qi}(V_i)}{V_i} dV_i \tag{4.135}$$

$$W_{24}(\theta,E'_q,V,\phi) = \sum_{i=n+1}^{2n} \left[E'^2_{qi} + V_i^2 - 2E'_{qi}V_i \cos(\theta_i - \phi_i) - (E'^2_{qio} + V_{io}^2 - 2E'_{qio}V_{io} \cos(\theta_{io} - \phi_{io}))\right] \frac{1}{2x'_{di}} \tag{4.136}$$

$$W_{25}(V,\phi) = -\frac{1}{2} \sum_{i=n+1}^{n+N} \sum_{j=n+1}^{n+N} B_{ij}(V_iV_j \cos \phi_{ij} - V_{io}V_{jo} \cos \phi_{ijo}) \tag{4.137}$$

$$W_{26}(\theta,V,\phi) = -\sum_{i=n+1}^{2n} \left[V_i^2 (\cos 2(\theta_i - \phi_i) - 1) - V_{io}^2 (\cos 2(\theta_{io} - \phi_{io}) - 1)\right] \frac{x'_{di} - x'_{qi}}{4x'_{di}x'_{qi}} \tag{4.138}$$

$$W_{27}(t) = -\sum_{i=1}^{n} \int_{t_o}^{t} \frac{E_{fdi}}{(x_{di} - x'_{di})} \frac{dE'_{qi}}{dt} dt \tag{4.139}$$

$$W_{28}(E'_q) = \sum_{i=1}^{n} \frac{E'^2_{qi} - E'^2_{qio}}{2(x_{di} - x'_{di})} \tag{4.140}$$

$$W_{29}(t) = \sum_{i=1}^{n} \int_{t_o}^{t} \frac{T'_{doi}}{(x_{di} - x'_{di})} \left(\frac{dE'_{qi}}{dt}\right)^2 dt \tag{4.141}$$

$$W_{2,10}(\theta,E'_d,V,\phi) = \sum_{i=n+1}^{2n} \left[E'^2_{di} + V_i^2 + 2E'_{di}V_i \sin(\theta_i-\phi_i) - \left(E'^2_{dio} + V_{io}^2 + 2E'_{dio}V_{io} \sin(\theta_{io}-\phi_{io})\right)\right] \frac{1}{2x'_{qi}} \tag{4.142}$$

$$W_{2,11}(V) = -\sum_{i=n+1}^{2n} \frac{V_i^2 - V_{io}^2}{2x'_{qi}} \tag{4.143}$$

$$W_{2,12}(t) = \sum_{i=1}^{n} \int_{t_o}^{t} \frac{T'_{qoi}}{(x_{qi} - x'_{qi})}\left(\frac{dE'_{di}}{dt}\right)^2 dt \tag{4.144}$$

$$W_{2,13}(E'_d) = \sum_{i=1}^{n} \frac{E'^2_{di} - E'^2_{dio}}{2(x_{qi}-x'_{qi})} \tag{4.145}$$

Subscript 'o' in the above expressions indicates the quantities at the post-fault s.e.p. In (4.136) and (4.142) E'_{qi} and E'_{di} pertain to the machines connected at that bus. It can be shown that the time derivative of the function defined in Eq. (4.131) is zero.

Comments

1) The last 4 terms $W_{2,10}$, $W_{2,11}$, $W_{2,12}$ and $W_{2,13}$ correspond to the change in energy due to the presence of damper winding on the quadrature axis. In the absence of this winding, $E'_{di} = 0$ and these terms vanish. Thus, there are only 9 terms in the energy function when one-axis model of the generator is used as in Padiyar and Sastry [1987]. The 9th term which is introduced makes the energy function term constant along the post-fault trajectory, i.e., $\dot{W} = 0$. If we did not use the 9th term, then $\dot{W}$ is given by

$$-\sum_{i=1}^{n} \frac{T'_{doi}}{(x_{di} - x'_{di})} \left(\frac{dE'_{qi}}{dt}\right)^2$$

which is non-positive.

2) The term W_{26} in Eq. (4.138) is the change in energy stored due to saliency in the absence of which the term is zero.

3) In the absence of the excitation system, E_{fd} is constant and W_{27} in Eq. (4.139) can be expressed as

$$-\sum_{i=1}^{n} E_{fdi} \frac{E'_{qi} - E'_{qio}}{x_{di} - x'_{di}} \tag{4.146}$$

4) W_{22}, the potential energy component due to active power loads, is path dependent when P_{Li} is time-varying due to voltage dependence, and has to be evaluated by numerical integration. However, the term can be approximated as

$$W_{22} = \sum_{i=1}^{n} \left[f_{pi}(V_i)\, \phi_i - f_{pi}(V_{io})\, \phi_{io}\right] \tag{4.147}$$

If P_{Li} is constant, W_{22} reduces to

$$W_{22} = \sum_{i=1}^{n} P_{Li}\, (\phi_i - \phi_{io}) \tag{4.148}$$

which is path independent.

5) The governor and prime mover control system has little effect on the first swing stability. However, if these effects are included, mechanical power will become time-varying and the term W_{21} modifies to

$$W_{21} = -\sum_{i=1}^{m} \int_{t_o}^{t} P_{mi}(t) \frac{d\theta_i}{dt}\, dt \tag{4.149}$$

Special case

6) If a fault at its terminals separates generator k from the rest of the generators, then the kinetic energy term can be modified to $\hat{W}_1$ defined as

$$\hat{W}_1 = \frac{1}{2} \frac{M_k M_{T-k}}{M_T} (\tilde{\omega}_k - \tilde{\omega}_{T-k})^2 \tag{4.150}$$

where
$$M_{T-k} = \sum_{\substack{i=1 \\ \neq k}}^{n} M_i , \qquad M_T = \sum_{i=1}^{n} M_i \tag{4.151}$$

$$\omega_{T-k} = \frac{1}{M_{T-k}} \sum_{\substack{i=1 \\ \neq k}}^{m} M_i \omega_i \tag{4.152}$$

The subscript T-k refers to the rest of the generators.

4.7 Lyapunov Function for Multimachine Power Systems with Flux Decay and Non-linear Voltage Dependent Loads (Bergen, Hill and DeMarco [1984])

In this section we consider the flux decay model with nonlinear voltage dependent loads. We adopt the same numbering scheme as in Chapter 2. We assume real power loads to be frequency dependent while the reactive power loads are voltage dependent. Thus,

$$\begin{aligned} P_{\ell i} &= P^o_{\ell i} + D_i \dot{\phi}_i & i &= n+1, \dots n+N \\ Q_{\ell i} &= Q_i (|V_i|) & i &= n+1, \dots n+N \end{aligned} \tag{4.153}$$

The differential equations are

$$M_i \ddot{\delta}_i + D_i \dot{\delta}_i = P_{mi} - P_{bi} (V) \tag{4.154}$$

$$T'_{doi} \frac{dE'_{qi}}{dt} = E_{fdi} - E'_{qi} - \frac{(x_{di} - x'_{di})}{x'_{di}} (E'_{qi} - |V_{i+n}| \cos \delta_m) \qquad i = 1, 2 \dots n \tag{4.155}$$

where δ_m is the phase angle between E'_{qi} and V_{i+n}, i.e., $\underline{/E'_{qi}} - \underline{/V_{i+n}}$. From Eq. (4.81), $I_{di} = (E'_{qi} - V_{qi})/x'_{di}$ and from Fig. 4.2, V_{qi} is the projection of V_{ti} (= V_{i+n}) on the q axis. Hence $V_{qi} = |V_{i+n}| \cos \delta_m$. Use of these facts in (4.74) yields Eq. (4.155). Multiplying both sides of (4.155) by $\left(\frac{1}{x_{di} - x'_{di}} \right)$ E'_{qi}, we get

$$\frac{T'_{doi}}{x_{di} - x'_{di}} E'_{qi} \frac{d}{dt} E'_{qi} = E'_{qi} \frac{(E_{fdi} - E'_{qi})}{(x_{di} - x'_{di})} - \left[\frac{E'^{2}_{qi}}{x'_{di}} - \frac{E'_{qi} |V_{i+n}| \cos \delta_m}{x'_{di}} \right]$$

$$i = 1, 2, \ldots n \qquad (4.156)$$

The term in the brackets can be viewed as the reactive power supplied by E'_{qi}, i.e., by the fictitious bus voltage E'_{qi} to a "transmission line" with series reactance x'_{di}. If we ignore the angle ϕ_i as in Sec. 4.4.2, then $E'_{qi} = V_i + j\, x'_{di} I_i$ and the equivalent circuit is given by Fig. 4.11.

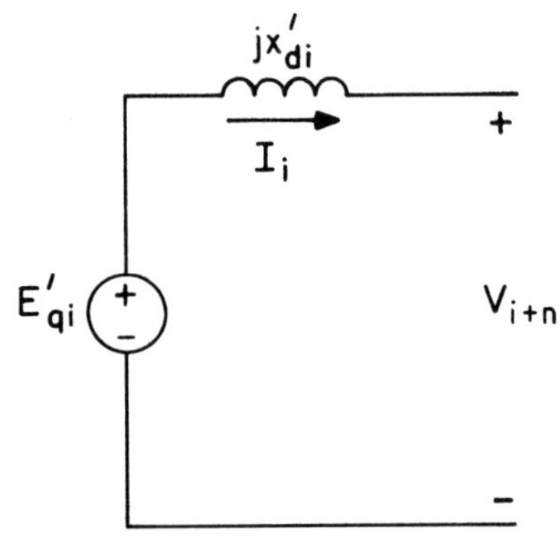

Figure 4.11 Equivalent circuit with flux decay model.

Replacing notationally E'_{qi} by $|V_i|$, we replace Eq. (4.156) by

$$\varepsilon_i |V_i||\dot{V}_i| + Q_{bi}(|V_i|) = Q_{Gi}(|V_i|) \tag{4.157}$$

where $\varepsilon_i = \dfrac{T'_{doi}}{x_d - x'_{di}} > 0$

$$|V_i| = E'_{qi} \tag{4.158}$$

$$Q_{Gi}(|V_i|) = \frac{E_{fdi}\ |V_i| - |V_i|^2}{x_{di} - x'_{di}} \tag{4.159}$$

$$Q_{bi}(|V_i|) = \frac{|V_i|^2}{x'_{di}} - \frac{|V_i||V_{i+n}|}{x'_{di}} \cos \delta_m \tag{4.160}$$

Note that Q_{Gi} is a fictitious reactive power supplied by E_{fdi} to the fictitious bus. It is <u>not</u> the actual reactive power supplied by the generator to the physical generator terminal.

The differential equations can also be written as

$$M_i \ddot{\delta}_i + D_i \dot{\delta}_i = P_{mi} - \sum_{j=1}^{n+N} |V_i||V_j| B_{ij} \sin \delta_{ij} \tag{4.161}$$

$$\varepsilon_i |V_i||\dot{V}_i| = Q_{Gi}(|V_i|) + \sum_{j=1}^{n+N} |V_i||V_j| B_{ij} \cos (\delta_i - \delta_j) \qquad i=1, 2 \ldots n \tag{4.162}$$

where B_{ij} is the susceptance part of the Y_{ij} term in the bus admittance matrix of the augmented system. Note that if the reactance between buses i and j is x_{ij}, then $B_{ij} = +\dfrac{1}{x_{ij}}$ and $B_{ii} < 0$. Hence, in the equation for $|\dot{V}_i|$ in (4.162)

$$\sum_{j=1}^{n+N} |V_i||V_j| B_{ij} \cos(\delta_i - \delta_j)$$

$$= -\frac{1}{x'_{di}} E'^{\,2}_{qi} + \frac{E'_{qi} |V_{i+n}| \cos \delta_m}{x'_{di}} \tag{4.163}$$

Hence, (4.162) is equivalent to (4.157).

The load flow equations are

$$P^o_{\ell i} + D_i \omega_i + \sum_{j=n+1}^{n+N} |V_i||V_j| B_{ij} \sin \delta_{ij} = 0 \tag{4.164}$$

$$i = n+1, \; \ldots \; n+N$$

$$Q_i(|V_i|) - \sum_{j=n+1}^{n+N} |V_i||V_j| B_{ij} \cos \delta_{ij} = 0 \tag{4.165}$$

The overall post-fault system is then given by combining (4.161) - (4.162) and (4.164) - (4.165).

$$M_i \ddot{\delta}_i + D_i \dot{\delta}_i + P_{bi}(|V|,\delta) = P_{mi} \tag{4.166}$$

$$i=1,\; 2 \; \ldots \; n+N$$

$$\varepsilon_i \; |V_i||\dot{V}_i| + Q_{bi}(|V|,\delta) = Q_i(|V|,\delta) \tag{4.167}$$

where P_{bi} and Q_{bi} (i=1, 2 ... n+N) are the total real and reactive powers leaving the ith bus via transmission lines, i.e.,

$$P_{bi}(|V|,\delta) = \sum_{j=1}^{n+N} |V_i||V_j| B_{ij} \sin \delta_{ij} \tag{4.168}$$

$$Q_{bi}(|V|,\delta) = -\sum_{j=1}^{n+N} |V_i||V_j| B_{ij} \cos \delta_{ij} \tag{4.169}$$

B_{ij} is the susceptance part of the Y_{ij} term in the bus admittance matrix of the augmented system. The vector $|V| = [\,|V_1| \,\ldots\ldots\, |V_{n+N}|\,]^T$

$M_i > 0$ $i=1, 2 \ldots n$

$M_i = 0$ $i=n+1, \ldots n+N$

$D_i > 0$ $i=1, 2 \ldots n$ (generator damping coefficient)

$D_i > 0$ $i=n+1, \ldots n+N$ (load frequency coefficient)

$$\varepsilon_i = \frac{T'_{doi}}{x_{di} - x'_{di}} > 0 \qquad i=1, 2 \ldots n$$

$\varepsilon_i = 0$ $i=n+1, \ldots n+N$

$P_i = P_{mi} - P_{\ell i}$ $i=1, 2 \ldots n+N$

$P_{\ell i} = 0$ $i=1, 2 \ldots n$

$P_{mi} = 0$ $i=n+1, \ldots n+N$

$$Q_i(|V_i|) = Q_{Gi}(|V_i|) - Q_{\ell i}(|V_i|,)$$

$$Q_{Gi}(|V_i|) = \frac{|V_i|\,(E_{fdi} - |V_i|)}{x_{di} - x'_{di}} \qquad i=1, 2 \ldots n \tag{4.170}$$

$Q_{\ell i} = 0$ $i=1, 2 \ldots n$

$Q_{Gi} = 0$ $i=n+1, \ldots n+N$

Note that $Q_{\ell i}$ $(i=n+1, \ldots n+N)$ and Q_{Gi} $(i=1, 2 \ldots N)$ depend only on $|V_i|$.

Let us make the assumption that $\sum_{i=1}^{n+N} (P_{mi} - P_{\ell i}) = 0$. The equilibrium is obtained by solving $\ddot{\delta} = \dot{\delta} = |\dot{V}| = 0$. Also $\dot{\phi}_i = 0$ $(i=n+1, \ldots n+N)$. At equilibrium we have

$$P_{bi}(|V|,\delta) = P_{mi} \qquad i=1, 2 \ldots n+N \tag{4.171}$$

$$Q_{bi}(|V|,\delta) = Q_i(|V_i|)$$

The two equations can be compactly represented as

$$S_{bi}(V) = S_i(|V_i|) \tag{4.172}$$

where the $\quad \mathrm{Re}\left[S_i(|V_i|)\right] = P_{mi}$

$$\mathrm{Im}\left[S_i(|V_i|) = Q_i(|V_i|)\right]$$

As in Chapter 2, define $\alpha_i = \delta_i - \delta_n$ $i=1, 2 \ldots n+N$ $(i \neq n)$. Instead of the $(n+N-1)$ dimensional vector α we define a n dimensional vector $\tilde{\alpha}$ with the same elements as α except for the $(n+N)^{th}$ element which is zero. This is done for convenience. Note that in our notation n^{th} bus is the reference bus which is also the slack generator internal bus.

$$\tilde{\alpha} = \tilde{T}\,\delta$$

$$\tilde{T} = \begin{bmatrix} I_{n-1} & -e_{n-1} \\ 0^T & 0 \end{bmatrix} \tag{4.173}$$

Introduce the internodal voltages

$$\nu_i = |V_i|\, e^{j\alpha_i} \qquad i=1, 2 \ldots n+N$$

Let $\nu = [\nu_1 \ldots \nu_{n+N}]^T$. Define $S_b = [S_{b1} \ldots S_{b(n+N)}]^T$ and $S = [S_1 \ldots S_{(n+N)}]^T$. From nodal analysis

$$S_b(V) = -j[V]\,BV^* = -j[\nu]\,B\nu^* \tag{4.174}$$

where $\quad B$ = susceptance matrix of order $(n+N)\times(n+N)$, $[V] = \mathrm{diag}\,(V_i)$ and V_i^* is the complex conjugate of V_i. Similar definitions hold for $[\nu]$ and ν^*.

The equations (4.166) and (4.167) are now written as

$$M\ddot{\delta} + D\dot{\delta} + \tilde{T}^{T} (P_b(\nu) - P_m) \qquad (4.175)$$

$$i=1, 2, \ldots n+N$$

$$E\,[|\nu|]\,|\dot{\nu}| + Q_b(|\nu|) - Q(|\nu|) = 0 \qquad (4.176)$$

$$[E] = \text{Diag}\,(\varepsilon_i)$$

where δ and ν are (n+N)-vectors, and M and D are diagonal (n+N) matrices. For brevity the explicit dependence of P_b and Q on $\tilde{\alpha}$ is not indicated. Also note that $\tilde{\alpha}$ is a (n+N) dimensional vector as defined in (4.173). The states chosen are the n angular velocities $\dot{\delta}_i = \omega_i$, (n+N) internodal angles α_i and the n+N voltage magnitudes $|\nu_i|$. Equations (4.175)-(4.176) are written in state space form as

$$\dot{\omega}_g = -M_g^{-1} D_g\, \omega_g - M_g^{-1}\, \tilde{T}_2^T\, (P_b(\nu) - P_m) \qquad (4.177a)$$

$$\dot{\tilde{\alpha}} = \tilde{T}_2\, \omega_g - \tilde{T}_1\, D_\ell^{-1}\, \tilde{T}_1^T\, (P_b(\nu) - P_m) \qquad (4.177b)$$

$$|\dot{\nu}| = -E^{-1}\, [|\nu|]^{-1}\, (Q_b(\nu) - Q(|\nu|)) \qquad (4.177c)$$

where the subscripts 'g' and 'ℓ' refer to generators and network, respectively. $\tilde{T}_1$ and $\tilde{T}_2$ are partitioned matrices of $\tilde{T}$ as in Chapter 2.

We now combine (4.177b) and (4.177c) into one set of complex vector differential equations as follows.

Consider $\nu_i = |V_i|\, e^{j\alpha_i}$ $\quad i=1, 2, \ldots n+N$ $\qquad$ (4.178)

Obviously $|V_i| = |\nu_i|$. Differentiating (4.178)

$$\dot{\nu}_i = |\dot{\nu}_i|\, e^{j\alpha_i} + |\nu_i|\, je^{j\alpha_i}\,\dot{\alpha}_i$$

$$= |\nu_i| \left[\frac{|\dot{\nu}_i|}{|\nu_i|} + j\dot{\alpha}_i \right] \tag{4.179}$$

For the nth component we note that (4.179) $\alpha_n \equiv 0$, implies $\dot{\alpha}_n = 0$.

The matrix notation (4.179) can be written as

$$\dot{\nu} = [\nu] \left([|\nu|]^{-1} |\dot{\nu}| + j\dot{\alpha} \right) \tag{4.180}$$

We now replace (4.177b) and (4.177c) by a single complex equation as

$$\dot{\nu} = [\nu] \left\{ [|\nu|]^{-1} [-E]^{-1} [|\nu|]^{-1} \left(Q_b(\nu) - Q(\nu) \right) + j\dot{\alpha} \right\}$$

$$= - [E]^{-1} [\nu] [|\nu|]^{-2} \left(Q_b(\nu) - Q(\nu) \right) + j\, [\nu]\, \tilde{T}_2\, \omega_g$$

$$- j\, [\nu]\, \tilde{T}_1\, D_\ell^{-1} T_1^T \left(P_b(\nu) - P_m \right) \tag{4.181}$$

Thus either (4.177a)-(4.177c) or (4.177a) and (4.181) can be used.

Using (4.181), the Lyapunov function is given by

$$V(\omega_g, \nu, \nu^\circ)$$

$$= \frac{1}{2}\, \omega_g^T\, M_g\, \omega_g + W(\nu,\nu^\circ) \tag{4.182}$$

where $W(\nu,\nu^\circ) = \text{Im} \int_{\nu^\circ}^{\nu} \langle [z]^{-1} (S_b(z) - S(|z|))\, dz \rangle$.

$\langle x,y \rangle$ denotes the product $x^T y$ in the (n-1) dimensional complex unitary space. (This is not an inner product.) Superscript o indicates equilibrium value.

Using (4.174), $W(\nu,\nu^\circ) = \text{Im} \int_{\nu^\circ}^{\nu} \langle (-jBz^* - [z]^{-1} S(|z|)) \, dz \rangle$

$$= -\text{Re} \int_{\nu^\circ}^{\nu} \langle Bz^*, dz \rangle - \text{Im} \int_{\nu^\circ}^{\nu} \sum_{i=1}^{n+N} \frac{S_i(|z_i|)}{z_i} dz_i \tag{4.183}$$

Now use the formula $Y_{BUS} = A\, y\, A^T$ where y is the primitive admittance matrix and A is a reduced incidence matrix for the augmented network. Hence, $B = A\, B_b\, A^T$ where B_b is a diagonal matrix whose elements are the branch admittances. Then the first term becomes

$$\text{Re} \int_{\nu^\circ}^{\nu} \langle Bz^*, dz \rangle = \text{Re} \int_{\nu^\circ}^{\nu} \langle B_b\, A^T z^*, d\, A^T z \rangle$$

$$= \sum_{i=1}^{\ell} \text{Re} \int_{\mu_i}^{\mu_i} B_{bi}\, \eta_i^*\, d\eta_i \quad \text{where } \eta = A^T z$$

$$\mu = A^T \nu$$

$$= \frac{1}{2} \sum_{i=1}^{\ell} B_{bi} \left(|\mu_i|^2 - |\mu_i^\circ|^2\right)$$

$$= \frac{1}{2} \langle \nu^*, B\nu \rangle - \frac{1}{2} \langle \nu^{\circ *}, B\nu^\circ \rangle \tag{4.184}$$

Similarly, the second term in (4.183) can be shown to be equal to

$$= \sum_{i=1}^{n+N-1} P_{mi}(\alpha_i - \alpha_i^\circ) + \sum_{i=1}^{n+N} \int_{|\nu_i^\circ|}^{|\nu_i|} \frac{Q_i(|z_i|)}{|z_i|} d|z_i| \tag{4.185}$$

Since α_n is zero the upper limit on the first term in (4.185) is (n+N-1).

Hence, $W(\nu,\nu^\circ)$

$$= -\frac{1}{2} \langle \nu^*, B\nu \rangle + \frac{1}{2} \langle \nu^{\circ *}, B\nu^\circ \rangle$$

$$- \sum_{i=1}^{n+N} \int_{|\nu_i^\circ|}^{|\nu_i|} \frac{Q_i(|z_i|)}{|z_i|} d|z_i| - \sum_{i=1}^{n+N-1} P_{mi}(\alpha_i - \alpha_i^\circ) \tag{4.186}$$

We can evaluate the terms more explicitly if desired. For example

$$< \nu^*, B\nu > = \sum_{i=1}^{n+N} \sum_{j=1}^{n+N} B_{ij} |\nu_i||\nu_j| \cos(\alpha_i - \alpha_j)$$

$$\sum_{i=1}^{n+N} \int_{|\nu_i^\circ|}^{|\nu_i|} \frac{Q(|z_i|)}{|z_i|} d|z_i| = - \sum_{i=n+1}^{n+N} \int_{|\nu_i^\circ|}^{|\nu_i|} \frac{Q_{\ell i}(|z_i|)}{|z_i|} d|Z_i|$$

$$+ \sum_{i=1}^{n} \int_{|\nu_i^\circ|}^{|\nu_i|} \frac{Q_{Gi}(|z_i|)}{|z_i|} d|z_i| \tag{4.187}$$

Also
$$\sum_{i=1}^{n+N-1} P_{mi}(\alpha_i - \alpha_i^\circ) = - \sum_{i=n+1}^{n+N} P_{\ell i}(\alpha_i - \alpha_i^\circ) + \sum_{i=1}^{n-1} P_{mi}(\alpha_i - \alpha_i^\circ) \tag{4.188}$$

We note that the second term on the right-hand side in (4.187) can be explicitly evaluated using (4.170) as

$$\sum_{i=1}^{n} \frac{E^\circ_{fdi}}{x_{di} - x'_{di}} (|\nu_i| - |\nu_i^\circ|) + \frac{|\nu_i^\circ|^2}{2} - \frac{|\nu_i|^2}{2} \tag{4.189}$$

It has been shown in Bergen, Hill, DeMarco (1984) that V is indeed positive definite by showing that the Hessian of the potential function $W(\nu,\nu^o)$ is positive definite. Define

$$x = \begin{bmatrix} |\nu| \\ \alpha \end{bmatrix}$$

From (4.186) and (4.187) it is clear that $W(\nu,\nu^o) = \hat{W}(|\nu|,\alpha) = \hat{W}(x)$. It has been shown that $\hat{W}(x)$ is locally positive definite in a neighborhood of the equilibrium point ν^o if $\frac{\partial^2 W}{\partial x^2}$ evaluated at $x = x_o$ is positive definite. They have also shown that this function by proper assumption can be reduced to the structure preserving energy function using classical models.

4.8 Conclusion

In this chapter some of the work in using energy functions for detailed models has been discussed. However, it still remains a big research area in terms of validation on practical systems. Except for flux decay models, the rest of the modeling details have to be accounted for by a numerical energy function. Some quasi-analytical energy functions are also discussed. Numerical results of the method of Sec. 4.6 on a 10 machine system are to be found in Padiyar and Ghosh (1987).

Chapter 5

REGION OF STABILITY IN POWER SYSTEMS

5.1 Introduction

The stability problem in power systems was mathematically formulated in Chapter I as one of ensuring that the state of the power system at the instant of clearing the fault is inside the region of stability (ROS) of the post-fault stable equilibrium point. Computation of ROS is perhaps the most difficult task in successfully using the Lyapunov/energy functions for stability analysis. In this chapter we explain the foundation of the theory underlying the characterization of the stability boundary of nonlinear autonomous dynamical systems and then indicate its application to power systems. We also provide a theoretical foundation to the potential energy boundary surface method. Both these theoretical results are due to Chiang et al. [1988]. A parallel development in characterizing ROS in the entire state space is due to Zaborsky et al. [1988]. In terms of application to realistic systems, three basic methods have proved to be successful in application.

1. The controlling u.e.p. method due to Athay, Podmore and Virmani [1979]
2. The potential energy boundary surface (PEBS) method due to Kakimoto, Ohsawa and Hayashi [1978]
3. Acceleration method due to Ribbens Pavella et al. [1979]

Good results for critical clearing times using these methods or their variations have been claimed by several researchers. These practical algorithms will be discussed in Chapter VI. In the case of the controlling u.e.p. method, corrections to both the potential and kinetic energy terms are necessary. In the case of the PEBS method, an accurate trajectory approximation is sought or the iterative PEBS method is used. Method 3 has been discussed earlier in Pai [1981] in detail. However, its most recent application to the 2-area energy function will be discussed in the next chapter.

5.2 Characterization of the Stability Boundary

The presentation in this section largely follows that in Chiang et al. [1987]. Consider the autonomous system

$$\dot{x} = f(x) \tag{5.1}$$

Definitions

1. A point x_i is called an equilibrium point (e.p.) of (5.1) if $f(x_i) = 0$.

2. If the Jacobian of f(x) at x_i denoted by $J(x_i)$ has no eigenvalues with zero real parts, then x_i is termed a hyperbolic e.p.

3. The type of an e.p. is defined by the number of eigenvalues of $J(x_i)$ in the right half plane. It is a type-one e.p. if the Jacobian has exactly one eigenvalue in the right half plane. Similarly we can define type-two e.p., etc.

Let E denote the set of e.p.'s and E_1 the set of type-one e.p.'s. Let x(t) denote the solution of (5.1)

Given an s.e.p. x_s there is a region in the state space from which trajectories converge to x_s. This is called the stability region of x_s and is denoted by $A(x_s)$, i.e.,

$$A(x_s) := \left\{ x \,\middle|\, \lim_{t \to \infty} x(t) = x_s \right\}$$

The stability boundary is denoted by $\partial A(x_s)$. Let x_i be a hyperbolic e.p. of (5.1). Its stable manifold $W^s(x_i)$ and unstable manifold $W^u(x_i)$ are defined as

$$W^s(x_i) = \{x \mid x(t) \to x_i \text{ as } t \to \infty\}$$
$$W^u(x_i) = \{x \mid x(t) \to x_i \text{ as } t \to -\infty\}$$

It is clear both the stable and unstable manifolds are invariant sets, i.e., if the initial condition of x is on the manifold, then the trajectory x(t) will always stay on that manifold. We can also define the region of stability of x_s as the stable manifold of x_s i.e., $W_s(x_s)$.

Let us assume that for (5.1) we have found a s.e.p. whose region of stability is to be found. Furthermore we make the following assumptions regarding (5.1).

(A1) All equilibrium points of (5.1) on the stability boundary are hyperbolic.

(A2) The intersection of $W^s(x_i)$ and $W^u(x_i)$ satisfies the transversality condition for all equilibrium points x_i and x_j on the stability boundary. Physically this implies that either the two manifolds do not intersect at all or if they do intersect then at the intersection point(s) the tangent hyperplanes span the state space. (Fig. 5.1)

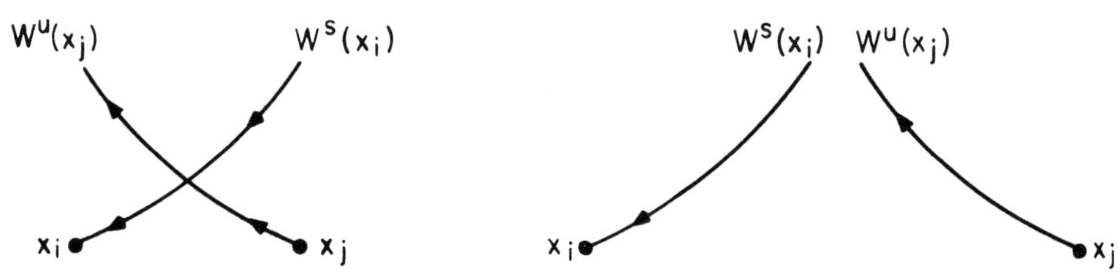

Fig. 5.1 Satisfaction of transversality condition

(A3) There exists a C^1 function $V(x)$: $R^n \to R$ for (5.1) such that

(i) $\dot{V}(x) \leq 0$ for $x \notin E$

(ii) If x is not an e.p., then the set $\{t \in R: \dot{V}(x)=0)\}$ has measure 0 in R. This is in fact equivalent to the LaSalle-Lefschetz condition in Lyapunov stability theory (Miller and Michel [1982]) which states that at points where $\dot{V}(x) = 0$ other than the e.p.'s, those points do not constitute a solution of (5.1)

(iii) V(x) is bounded implies that the x(t) is bounded.

Remarks: (1) In Lyapunov theory we insist on V(x) being p.d. whereas here we only require V(x) to belong to class C^1 and bounded.

Next we state without proof two important theorems concerning stability regions.

Theorem 1: (Necessary and sufficient condition for a u.e.p. to be on the stability boundary)

For the system (5.1) satisfying assumptions (A1)-(A3), x_i is an unstable e.p. on the stability boundary $\partial A(x_s)$ of a stable e.p. if and only if $W^u(x_i) \cap A(x_s) \neq \phi$.

Theorem 2: (Characterization of the stability boundary)

For the dynamical system (5.1) satisfying assumptions (A1)-(A3) let x_i (i=1, 2, ...) be the u.e.p's on the stability boundary $\partial A(x_s)$ of the s.e.p x_s. Then $\partial A(x_s) = \bigcup_{x_i \in E \cap \partial A} W^s(x_i)$

Remarks: 1. For proofs of these theorems refer to Chiang et al. [1987].

2. Theorem 2 is a generalization of an earlier theorem due to Tsolas,

Araposthasis and Varaiya [1985]. They conjectured that the e.p's on the stability boundary are of type one. Using theorems (1) and (2) it is possible theoretically to construct the stability boundary by obtaining analytical expressions for the stable manifolds of e.p's lying on the stability boundary. However, except for planar systems these constructive procedures are of little use. We now relate the stability boundary to the energy function approach and try to characterize the stability region locally near the point where the trajectory crosses the ROS when the system is cleared.

Theorem 3: (Energy function and the equilibrium points)

If the dynamical system (5.1) satisfies the assumption (A3), then on the stable manifold $W^s(x_i)$ of an e.p x_i, $V(x)$ achieves the minimum at $x = x_i$ i.e., $V(x_i) = \min V(x)$ for $x \in W^s(x_i)$

We will now apply the theorems to a classical power system model. Using the internal node model, let the terminal of the $(n+1)^{th}$ generator be an infinite bus. All angles of the other generators are measured with respect to this terminal of the $(n+1)^{th}$ bus. The angle of this terminal is constant and taken as zero. Hence, $\omega_{n+1} = \bar{\omega}$ = constant. The dynamics of the n machines are now described by (5.2), where the transfer conductances are neglected

$$\dot{\delta}_i = \omega_i \qquad i = 1, 2, \ldots n$$

$$M_i\dot{\omega}_i = P_i - D_i\omega_i - \sum_{\substack{j=1 \\ \neq i}}^{n+1} C_{ij} \sin \delta_{ij} \tag{5.2}$$

where $C_{ij} = |E_i||E_j| B_{ij}$, ω_i is the relative rotor velocity between the ith machine and $\bar{\omega}$. Note that through this formulation we are in the minimal state space. If the $(n+1)^{th}$ bus was not an infinite bus then there is need for an additional equation. Since its velocity is not constant, the state variables then are the rotor angles of the n machines with respect to the $(n+1)^{th}$ machine and the rotor velocities of the (n+1) machines. Reverting to (5.2) denote the vectors $\delta = (\delta_1 \ldots \delta_n)^T$, $\omega = (\omega_1 \ldots \omega_n)^T$, $P = (P_1 \ldots P_n)^T$, $|E| = (|E_1|, \ldots |E_n|)^T$, $f = (f_1 \ldots f_n)^T$, $M = \text{diag}(M_1 \ldots M_n)$, $D = \text{Diag}(D_1 \ldots D_n)$. Equation (5.2) can be put in the form

$$\dot{\delta} = \omega$$

$$\dot{\omega} = P - D\,\omega - f(\delta, |E|) \tag{5.3}$$

The equations for the pre-fault, faulted and post-fault states differ only in terms of the parameters B_{ij}. For stability we consider (5.3) with the post-fault parameters.

For this model it has been shown in Pai [1981] that with $D \equiv 0$, it is a conservative system and has a well-defined Lyapunov function. With $D \not\equiv 0$, we make the assumption that the Jacobian matrix $J = (\frac{\partial f}{\partial \delta})$ is non-singular at the equilibrium points.

With this it can be shown that (i) the system (5.3) has only hyperbolic equilibrium points (assumption A1) and (ii) there exists an energy function V(x) satisfying assumption (A3) of the type

$$V(\delta, \omega) = \frac{1}{2} \sum_{i=1}^{n} M_i \omega_i^2 - \sum_{i=1}^{n} P_i (\delta_i - \delta_i^s)$$

$$- \sum_{i=1}^{n} \sum_{j=i+1}^{n+1} C_{ij} (\cos \delta_{ij} - \cos \delta_{ij}^s) \tag{5.4}$$

where $x^s = (\delta^s, 0)$ is the stable e.p. An interesting result concerning the stability region is the following: The stability region $A(x_s)$ of a stable equilibrium point for the classical model is unbounded.

5.3 Region of Stability

Closest u.e.p. method

Let $V(x)$ be an energy function for the system (5.3). Let $V^1_{cr} = \mathrm{Min}(V(x_i)$: $x_i \in E \cap \partial A(x_s))$. The equilibrium point where this minimum is achieved is denoted by x^u_1 and it is a type one e.p. It is called the closest u.e.p. in the literature. The value of V at x^u_1 is denoted V^1_{cr} and is called the critical energy of the closest u.e.p. method. Also, for all x such that $V(x) < V^1_{cr}$, x is inside the stability region. The region defined by $V(x) < V^1_{cr}$ may be an open region. The algorithm to find x^u_1 is as follows.

Algorithm

1. Find all type one e.p.'s
2. Order these e.p.'s according to the values of the energy function at these e.p.'s
3. Test these e.p.'s to see if they lie on the stability boundary. The test involves using some actual integration around these e.p.'s (see next section)
4. From Step 3, among the e.p.'s that lie on the stability boundary choose the one with minimum value of $V(x)$. This is x^u_1 and $V(x)$ evaluated at this point is V^1_{cr} of the closest u.e.p.
5. Using the fault-on trajectory compute $V(x)$.

To find stability

If at $t = t_{c\ell}$ on the faulted trajectory, if $V(x(t_{c\ell})) < V^1_{cr}$ then the system is stable.

To find critical clearing time

Find t at which $V(x) = V^1_{cr}$ on the faulted trajectory. This time t is t^1_{cr}. As shown later this is generally a conservative estimate of actual t_{cr} (i.e., $t^1_{cr} < t^{act}_{cr}$.

Remarks

The algorithm spelled out above differs from the one commonly cited in the literature in two respects.

(i) To determine closest u.e.p., the minimum over all u.e.p.'s on the stability boundary, i.e., over the set $E \cap \partial A(x_s)$ is used rather than the set of all u.e.p.'s, i.e., the set E. Note that if there is an u.e.p. not on the stability boundary but whose energy function V(x) is > $V(x_s)$ but smaller than the ones on the stability boundary then it will result in a stability region which is much smaller than that by the closest u.e.p. method. Hence, the importance of identifying the u.e.p.'s lying on the stability boundary (see Fig. 5.2).

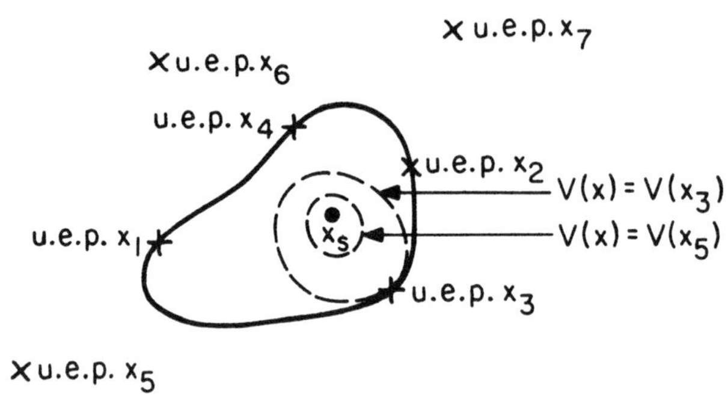

Figure 5.2 The u.e.p. x_3 gives minimum of V(x) on the stability boundary. The u.e.p. is not on the stability boundary and $V(x_5) < V(x_3)$. The estimation of stability region using $V(x_5)$ as critical energy is more conservative. (Reproduced from Chiang et al. [1987])

(ii) Algorithm to test if an u.e.p. is on the stability boundary

We assume that the e.p.'s lying on the stability boundary are of type one.

(i) Find the Jacobian at the type one e.p. (say $\hat{x}$).

(ii) Find the normalized unstable eigen vector. Note that there is only one such vector for a type one e.p.

(iii) Find the intersections of this unstable eigen vector with an ε-ball of the e.p. (The intersection points are $\hat{x} + \varepsilon y$ and $\hat{x} - \varepsilon y$.)

(iv) Integrate the post-fault system backward (reverse time) for some time from these intersection points. If the trajectory remains inside the ε-ball, go to the next step. Otherwise replace ε by $\alpha\varepsilon$ and also the intersection points $\hat{x} \pm \varepsilon y$ by $\hat{x} \pm \alpha\varepsilon y$ where $0 < \alpha < 1$. Repeat this step.

(v) Numerically integrate the equations starting from the intersection points.

(vi) If any of these trajectories approaches x_s, then the e.p. is on the stability boundary.

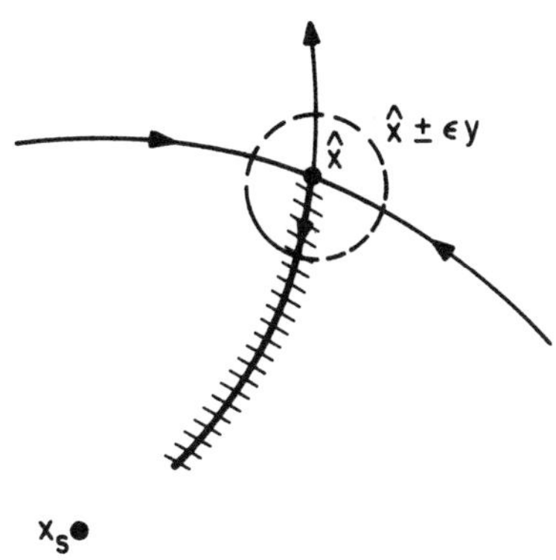

Figure 5.3 - Stable and unstable manifolds

In Steps (iv) - (vi) we basically test if the unstable manifold of a type one e.p. converges to the s.e.p. (hatched trajectory in Fig. 5.3).

5.3.1 Controlling u.e.p. method

Consider the fault on trajectory x(t) starting from x_0 at the time of the fault. Let the trajectory cross the region of stability at a point x_e called the exit point. The value of the energy function at the exit point is actually the true critical energy. Suppose x_e belongs to the stable manifold of $W^s(x_r^u)$ of the u.e.p. x_r^u on the stability boundary. Obviously $V(x_0) < V(x_r^u)$ and generally $V(x_1^u) < V(x_r^u)$. Hence, $V(x_r^u)$ is a better estimate of the true critical energy than the lowest u.e.p. method. We call x_r^u the relevant or the controlling u.e.p. Note that the controlling u.e.p. is not necessarily a type one u.e.p. Also the trajectory must pass the constant energy surface $\{x: V(x) = V(x_r^u)\}$ before it reaches the exit point x_e. If the system is conservative then on the stable manifold of x_r^u, $V(x^e) = V(x_r^u)$. In general $V(x^e)$ is greater than $V(x_r^u)$.

If from the fault-on trajectory we can determine the controlling u.e.p. then we have a better estimate of the critical energy. Conceptually the first two steps in the previous algorithm are modified as follows.

A1 From the fault-on trajectory find the controlling u.e.p. x_r^u.

A2 The critical energy V_{cr} is given by $V(x_r^u)$.

The determination of stability or t_{cr} is done as before.

To find the controlling u.e.p. x_r^u is not easy. Theoretically one may derive the stable manifolds of all the u.e.p.'s on the stability boundary. The controlling u.e.p. is the one whose manifold intersects with the fault-on trajectory first. By expressing the stable manifold in a power series expansion around the controlling u.e.p., we can use the same procedure as in the case of type one u.e.p. to ascertain if the u.e.p. belongs to the stability boundary.

The above steps for finding the controlling u.e.p. method are at best "conceptual" and no literature exists which has implemented the method. To make matters simple a conjecture in Tsolas et al. [1985] asserts that "almost all" exit points lie on the stable manifold of type one u.e.p.'s. This simplifies the procedure as follows.

A1 From the fault-on trajectory find the controlling type-one u.e.p. x^u_{r1}.

Algorithmically the steps are as follows

1. Derive the stable manifolds of all type one u.e.p.'s.

2. Determine the type-one u.e.p. the closure of whose manifold intersects with the fault-on trajectory first. Then that type-one u.e.p. is the controlling u.e.p.

Effort involved in Steps 1 and 2 may be simplified by deriving the tangent hyperplanes passing through all type-one u.e.p.'s on the stability boundary and then deriving the stable manifolds only for those type-one u.e.p.'s whose tangent hyperplanes intersect with the fault-on trajectory. A method of using the idea of using tangent hyperplanes due to Yee and Spalding [59] is discussed in the next section. The phase portrait of trajectory close to a type-one u.e.p. is shown in Fig. 5.4.

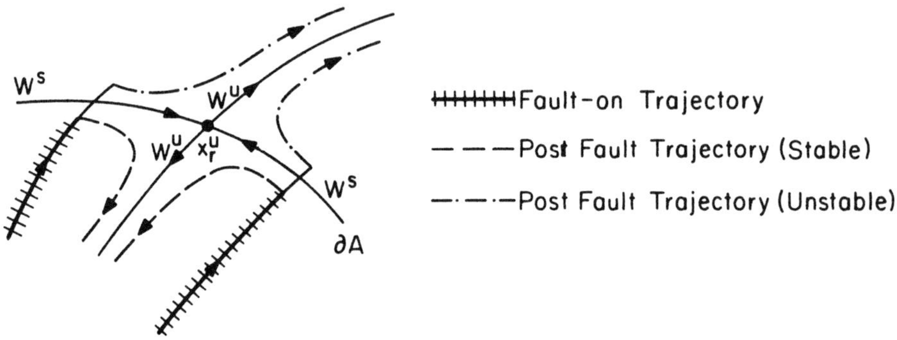

Figure 5.4 Trajectories of the faulted and post-fault system close to a controlling type-one u.e.p. x_r^u for different fault-clearing time.

Remarks: The preceding theoretical discussion is the firm foundation for computing the region of stability of the post-fault e.p. However, in systems greater than a single machine infinite bus system there are no numerical examples to back the theory and hence it offers an excellent area of further research.

For motivational purposes we illustrate the above concepts on a single machine infinite bus which is adapted from Chiang, Hirsch and Wu [1988].

Example 5.1: The following system of equations represent the swing equations of a classical model with saliency for a single machine infinite bus $x_1 = \delta$ and $x_2 = \omega$, the relative angular velocity.

$$\dot{x}_1 = x_2 \tag{5.5}$$

$$\dot{x}_2 = 0.301 - \sin(x_1 + 0.4136) + 0.138 \sin 2(x_1 + 0.4136) - 0.279x_2$$

The equilibrium points of (5.5) are periodic on the subspace $\{(x_1,x_2) \mid x_2 = 0\}$ and there exists a V-function

$$V(x_1,x_2) = 0.5x_2^2 + 0.301x_1 - \cos(x_1 + 0.4136) + 0.069 \cos 2(x_1 + 0.4136) \tag{5.6}$$

We have

$$\dot{V}(x_1,x_2) = \frac{\partial V}{\partial x_1}\dot{x}_1 + \frac{\partial V}{\partial x_2}\dot{x}_2 \tag{5.7}$$

$$= -0.279x_2^2 \leq 0$$

The Jacobian matrix of (5.5) at (x_1,x_2) is

$$J(x) = \begin{bmatrix} 0 & 1 \\ a & -0.279 \end{bmatrix} \tag{5.6}$$

where $a = -\cos(x_1 + 0.4136) + 0.276 \cos 2(x_1 + 0.4136)$

Let λ_1, λ_2 be the eigenvalues of $J(x)$. Then

$$\lambda_1 + \lambda_2 = -0.279 \tag{5.7a}$$

$$\lambda_1\lambda_2 = -a \tag{5.7b}$$

The following observations are immediate

(1) At least one of the eigenvalues must be negative which implies there is no source in the system (5.5). Hence as shown in Chiang et al. [1988] that the stability region (with respect to any stable equilibrium point) is unbounded.

(2) The stable equilibrium points and the type-one equilibrium points are located alternately on the x_1-axis.

It can be shown that (6.284098.0.0) is a stable equilibrium point of (5.5). Let us consider its stability region. The application of Theorem 1 shows that the type-one equilibrium points (2.488345.0.0) and (8.772443.0.0) are on the stability boundary. The stability region is again unbounded owing to the absence of a source. The stability boundary obtained by the present method is the curve B shown in Fig. 5.5 which is the union of stable manifolds of the equilibrium points (2.488345.0.0) and (8.772443.0.0). Curve A is the stability boundary obtained in Michel et al. [1982] (after a shift in coordinates). It is clear from the phase portrait in Fig. 5.6 that all the points inside curve B converge to the stable equilibrium point which verifies that the curve B is the exact stability boundary.

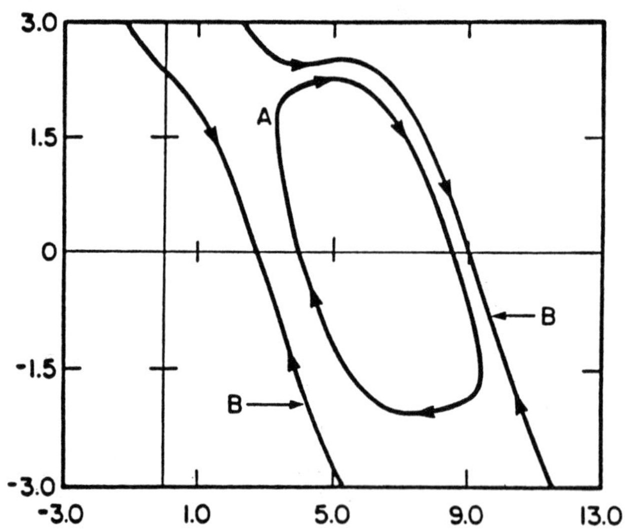

Fig. 5.5. Predictions of the stability region of Example 5.1 by different methods. (Reproduced from Chiang et al. [1988])

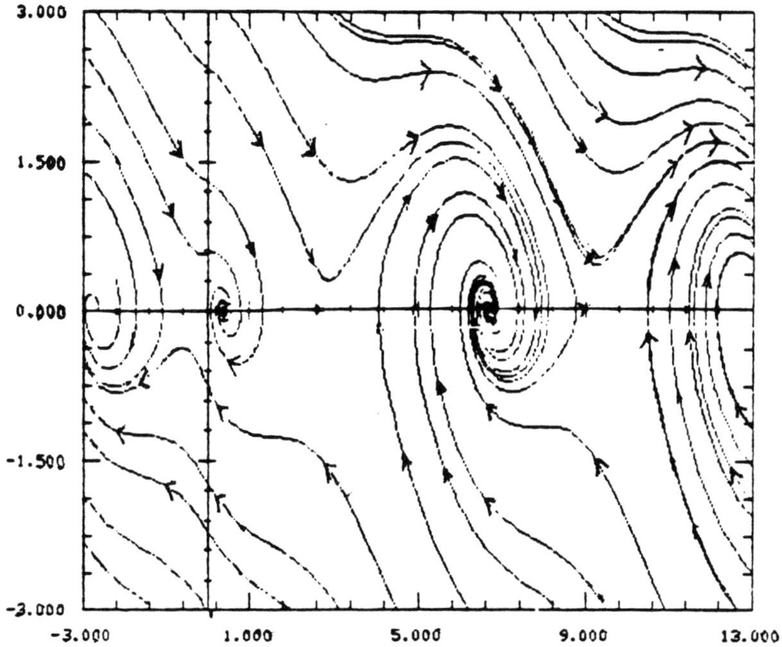

Fig. 5.6. Phase portrait of the system 5.5.
(Reproduced from Chiang et al. [1988])

5.4 Method of Hyperplanes and Hypersurfaces

5.4.1 Single machine system

Hyperplanes and hypersurfaces are special cases of stable manifolds and can be computed analytically. The work in this area is due to Yee [1975], Yee and Spalding [1977], Yee and Muir [1980], and Cook and Eskicioglu [1983]. The basic idea of the hyperplane and hypersurface method is illustrated through a single machine example.

For the post-fault system, the equations are

$$\begin{aligned} \dot{\delta} &= \omega \\ \dot{\omega} &= \frac{1}{M}\left[P_m - D\,\omega - P_e(\delta)\right] \end{aligned} \tag{5.8}$$

$\delta^u = \pi - \delta^s$ is the controlling unstable equilibrium point. The u.e.p. δ^u is transferred to the origin through the transformation

$$x = \delta - \delta^u \tag{5.9}$$

Thus, Eq. (5.8) becomes

$$\begin{aligned} \dot{x} &= \omega \\ \dot{\omega} &= \frac{1}{M}\left[P_m - D\omega - P_e(x + \delta^u)\right] \end{aligned} \tag{5.10}$$

The stability boundary is shown in Fig. 5.7 in the δ,ω plane.

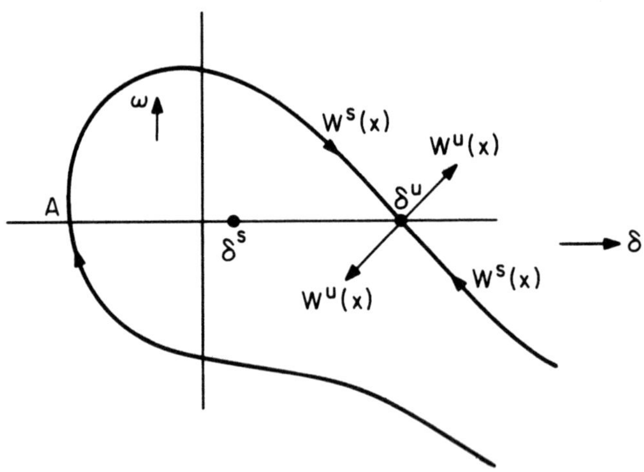

Figure 5.7 Stability boundary for a single machine system with damping.

It is seen that through the saddle point we have the stable and unstable manifolds with the union of the former constituting the stability boundary. We now seek an approximation of $W^s(x)$ around $x = 0$, i.e., around $\delta = \delta^u$. There are two forms possible

1) Explicit relationship of the type

$$\omega = \omega(x) \tag{5.11}$$

2) Implicit relationship $f(x,\omega) = 0$.

A disadvantage of (5.11) is the fact that it is single valued and valid till the point A in Fig. 5.7. But it is adequate for our purposes and moreover the algebra is simple.

1) <u>Explicit relationship</u>

Differentiating (5.11)

$$\dot{\omega} = \frac{\partial \omega}{\partial x} \dot{x} \tag{5.12}$$

Substituting in (5.10)

$$P_m - D\omega - P_e(x + \delta^u) = M \frac{\partial \omega}{\partial x} \dot{x} \tag{5.13}$$

Assume

$$\omega(x) = \sum_{n=1}^{\infty} \frac{1}{n!} \alpha_n x^n \tag{5.14}$$

where

$$\alpha_n = \left. \frac{d^n \omega}{dx^n} \right|_{x=0}$$

Substitution of (5.14) into (5.13) and equating coefficients of powers of x, α_n's, (n=1, 2 ...) are obtained to any degree desired. The details are in Yee [1975] and are omitted here. The results are shown in Fig. 5.8 for n=1, 2, 3, and 4. n=1 is the hyper plane and others are hyper surfaces.

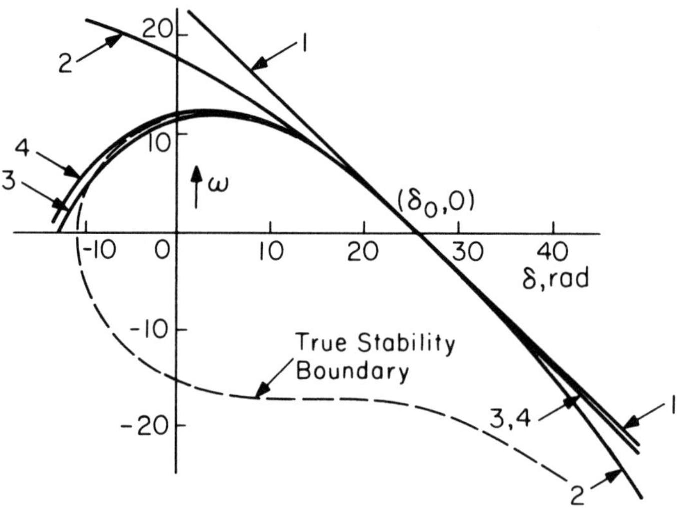

Figure 5.8 Power series representation of stability boundary. 1, 2, 3, 4 refer to the number of terms used. D = 10 p.u. (From Yee [1975])

2) Implicit relationship $f(x,\omega) = 0$ with $D = 0$

From (5.13)

$$\omega \frac{d\omega}{dx} = \frac{1}{M} [P_m - P_e (x + \delta^u)] \tag{5.15}$$

Integrating (5.15)

$$\frac{1}{2} \omega^2 = \frac{1}{M} [P_m x - \int_0^x P_e (x + \delta^u) dx] \equiv H(x) \tag{5.16}$$

(5.16) is indeed the stability boundary with $D = 0$. With $D \neq 0$, the stability boundary is assumed to be of the form

$$\frac{1}{2} \omega^2 = H(x) + \sum_{n=1}^{\infty} \frac{1}{(n+1)!} \beta_n x^{n+1} \tag{5.17}$$

Only two values of ω have significance namely for positive ω when $x < 0$ and for negative ω when $x > 0$, i.e., to the left and right of the saddle point, respectively. Differentiate (5.17) with respect to x

$$\omega \frac{d\omega}{dx} = \frac{dH}{dx} + \sum_{n=1}^{\infty} \frac{1}{n!} \beta_n x^n \tag{5.18}$$

From (5.10)

$$\frac{d\omega}{dx} = \frac{1}{M\omega} \left[P_m - D\omega - P_e (x + \delta^u)\right]$$

$$\omega \frac{d\omega}{dx} = \frac{1}{M} \left[P_m - P_e (x + \delta^u) - D\omega\right]$$

From (5.16)

$$\frac{dH}{dx} = \frac{1}{M} \left[P_m - P_e (x + \delta^u)\right]$$

Therefore

$$\omega \frac{d\omega}{dx} = \frac{dH}{dx} - \frac{1}{M} D\omega \tag{5.19}$$

Therefore comparing (5.18) and (5.19)

$$\sum_{n=1}^{\infty} \frac{1}{n!} \beta_n x^n = -\frac{1}{M} D\omega \tag{5.20}$$

ω is given by (5.14) as $\omega(x) = \sum_{n=1}^{\infty} \frac{1}{n!} \alpha_n x^n$.

Therefore, β_n $(n \geq 1)$ can be expressed in terms of α_n $(n \geq 1)$. Stability boundary using (5.17) converges more rapidly and is shown in Fig. 5.9.

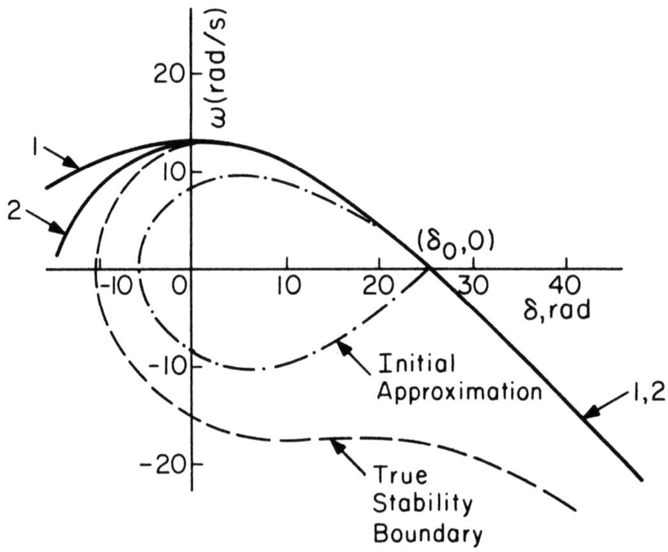

Figure 5.9 Stability boundaries with 1 and 2 terms in series using implicit relationship. (From Yee [1975])

The critical clearing times obtained by both the methods are quite accurate. We now discuss the extension of this approach to multimachine systems.

5.4.2 Multi-machine system

System differential equations

The treatment in this section largely follows that in Yee and Spalding [1977]. Each machine is represented by a constant voltage behind transient reactance, and loads are represented as constant impedances. Following reduction of the system to retain only the internal machine voltage nodes, the swing equation for the ith machine can be written as:

$$M_i\dot{\omega}_i = P_{mi} - E_i^2 G_{ii} - D_i\omega_i - \sum_{\substack{j=1 \\ j\neq i}}^{n} A_{ij} \cos(\delta_i - \delta_j - \phi_{ij}) \tag{5.21}$$

where $A_{ij} = |E_i||E_j||Y_{ij}|$ according to our earlier notation.

Assume uniform damping, i.e.,

$$\frac{D_i}{M_i} = C, \text{ for } i = 1 \text{ to } n.$$

The system is described by n-1 relative angles and n-1 relative speeds: Taking the nth machine as reference and defining:

$$\delta_{in} = \delta_i - \delta_n, \; \omega_{in} = \omega_i - \omega_n$$

a total of 2(n-1) equations result, namely

$$\begin{aligned} \dot{\delta} &= \omega \\ \dot{\omega} &= f(\delta) - C\omega \end{aligned} \tag{5.22}$$

where δ and ω are (n-1) vectors of the relative angles δ_{in} and relative speeds ω_{in}, respectively, and f is an (n-1) vector function with elements,

$$\begin{aligned} f_i(\delta) = {} & \frac{1}{M_i}(P_{mi} - E_i^2 G_{ii}) - \frac{1}{M_n}(P_{mn} - E_n^2 G_{nn}) \\ & - \frac{1}{M_i}\Big[A_{in}\cos(\delta_{in} - \phi_{in}) \\ & + \sum_{\substack{j=1 \\ j\neq i}}^{n-1} A_{ij}\cos(\delta_{in} - \delta_{jn} - \phi_{ij})\Big] \\ & + \frac{1}{M_n}\Big[\sum_{j=1}^{n-1} A_{jn}\cos(\delta_{jn} + \phi_{jn})\Big] \end{aligned} \tag{5.23}$$

The equilibrium points are the solutions of $\dot{\delta} = \dot{\omega} = 0$, i.e., $\omega = 0$ and $f(\delta) = 0$. We will assume that these are somehow computed. Linearizing (5.22) around an equilibrium point gives

$$\begin{bmatrix} \Delta\dot{\delta} \\ \\ \Delta\dot{\omega} \end{bmatrix} = \left[\begin{array}{cc} 0 & U \\ \hline J & -CU \end{array}\right] \begin{bmatrix} \Delta\delta \\ \\ \Delta\omega \end{bmatrix} = [K] \begin{bmatrix} \Delta\delta \\ \\ \Delta\omega \end{bmatrix} \tag{5.24}$$

where $J = \dfrac{\partial f(\delta)}{\partial \delta}$, U is a unity matrix and the elements of J are evaluated at the equilibrium point of interest.

The elements of J are

$$J_{ii} = \frac{1}{M_i}\Big[A_{in}\sin(\delta_{in} - \phi_{in}) + \sum_{\substack{j=1 \\ j\neq i}}^{n-1} A_{ij}\sin(\delta_{in} - \delta_{jn} - \phi_{ij})\Big] - \frac{1}{M_n} A_{in}\sin(\delta_{in} + \phi_{in}) \tag{5.25}$$

and the off diagonal elements are

$$J_{ij} = -\frac{1}{M_i} A_{in}\sin(\delta_{in} - \delta_{jn} - \phi_{ij}) - \frac{1}{M_n} A_{jn}\sin(\delta_{jn} + \phi_{jn}) \tag{5.26}$$

If the Newton's method is used to solve for $f(\delta) = 0$, then J is the Jacobian. The eigenvalues of K corresponding to each equilibrium point are of interest. If there is one eigenvalue with the real part greater than zero, then it is the type-one equilibrium point. If all eigenvalues have the real part less than zero, then it is a stable equilibrium point. Type one e.p.'s are important because the stability boundary passes through them as shown in the earlier section. The hyperplanes at each type-one equilibrium point give a local approximation for the separatrix or the stable and unstable manifolds at the equilibrium point.

The number of unstable equilibrium points which exist for a particular system is of interest. The formula $2^{n-1}-1$ has been suggested, but the true situation appears to be rather more complex. Experience has shown that the number of type-one points, in particular, is of the order of, and usually less than, the number of machines. Typical results are shown in Table 5.1.

Table 5.1: Number of type-one equilibrium points found for particular systems. (From Yee and Spalding [1977])

Number of machines	Number of type one points
3	2
5	5
6	6
15	10
25	20

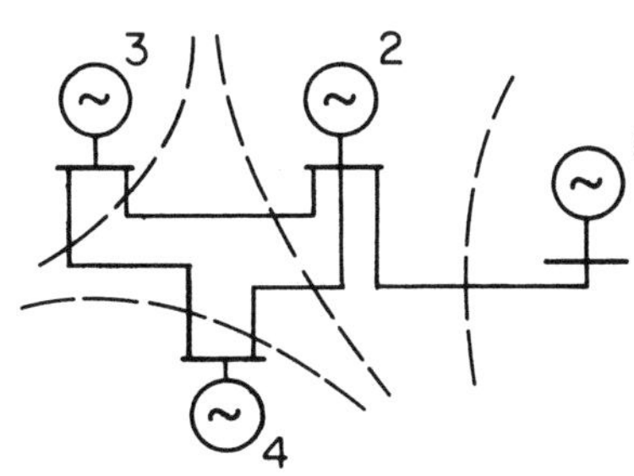

Fig. 5.10 Possible modes of separation of a typical 4-machine system.

The type one equilibrium points are, in fact, "non circulatory" in the terminology of Tavora and Smith [1972], i.e., each one corresponds to the system splitting into two asynchronous parts in a certain way. The ways in which a system can conceivably split are largely determined by the topology of the transmission network. This is a useful starting point in any search for type-one points. For example, for the four machine system shown in Fig. 5.10, there appear to be four possibilities, with the machines grouped as follows:

(1), (2,3,4)
(1,2), (3,4)
(1,2,4), (3)
or (1,2,3), (4).

However, it will be found for typical systems that equilibrium points do not necessarily exist corresponding to each of these possibilities; this further reduces the number of possible type one points. Furthermore, it has been found that not all points which describe the system separating into two parts are necessarily type-one equilibrium points.

Numerical methods for computing unstable equilibrium points have been widely discussed in the literature and will not be covered here. However, a fast method of determining the number of positive eigenvalues is desirable and this is discussed next.

Relationship between eigenvalues of K and J

Suppose λ is an eigenvalue of K with the associated eigen vector $(x_1^T, x_2^T)^T$ where x_1 and x_2 are (n-1) dimensional vectors, then we have

$$\left[\begin{array}{c|c} 0 & U \\ \hline J & -CU \end{array}\right] \begin{bmatrix} x_1 \\ x_2 \end{bmatrix} = \lambda \begin{bmatrix} x_1 \\ x_2 \end{bmatrix} \tag{5.27}$$

This gives

$$\begin{aligned} x_2 &= \lambda \; x_1 \\ Jx_1 &= \lambda \; (\lambda + C) \; x_1 \end{aligned} \tag{5.28}$$

Hence it follows that if λ' is an eigenvalue of J, then there are two corresponding eigenvalues of K such that

$$\lambda' = \lambda \; (\lambda + C)$$

$$\text{i.e.} \qquad \lambda = \frac{-\; C \pm \sqrt{C^2 + 4\;\lambda'}}{2} \tag{5.29}$$

The following possibilities exist

	λ'(Eigenvalue of J)	λ(Eigenvalue of K)
a)	positive real	positive real and negative real
b)	negative real	two negative real or a complex conjugate pair with negative real part equal to $-C/2$
c)	complex pair	two complex pairs

The eigenvalues of J are real if the transfer conductances are zero (i.e., $\phi_{ij} = \pi/2$)(cases (a) and (b)). Thus K and J have the same number of positive (real) eigenvalues and classification of the singular point can be based on either K or J. Case (c) will occur if there are non-zero transfer conductances. From this discussion it is clear that for a stable equilibrium point the eigenvalues of J are negative real and from the discussion in the literature it appears that when J has one positive real eigenvalue, the same is true for K.

Tangent hyperplane at type-one equilibrium point (Yee [1975])

To determine the equation for the hyperplane at a type one unstable equilibrium point we follow the procedure below. Let λ_1 be the eigenvalue of K with a positive real part. K^T has the same eigenvalue as K but different eigenvectors. Let y_1 be the eigenvector of K^T corresponding to λ_1. Then v_1 is orthogonal to the eigenvectors corresponding to the eigenvalues, namely $\lambda_2 \ldots \lambda_{2n-2}$ of K. Therefore

$$x_i^T y_1 = 0 \quad , \quad i = 2, 3, \ldots 2n-2 \tag{5.30}$$

i.e., vector y_1 is orthogonal to each of the eigenvectors $x_2, \ldots x_{2n-2}$ corresponding to the eigenvalues $\lambda_2 \ldots \lambda_n$ of K which are a basis for the hyper-plane. Hence, y_1 is normal to the hyperplane. Therefore, the equation for the hyperplane is

$$x^T y_1 = 0 \tag{5.31}$$

where x is any vector in the hyperplane. Note that x in (5.31) is the state vector $x = [\Delta\delta, \Delta\omega]^T$. If the nth element of y_1 is set to unity then the other elements of y_1 are obtained by solving

$$K^T \begin{bmatrix} \alpha_1 \\ \cdot \\ \cdot \\ \cdot \\ \alpha_{2n-3} \\ 1 \end{bmatrix} = \lambda_1 \begin{bmatrix} \alpha_1 \\ \cdot \\ \cdot \\ \cdot \\ \alpha_{2n-3} \\ 1 \end{bmatrix} = \lambda_1 \begin{bmatrix} \alpha \\ --- \\ \beta \end{bmatrix} \tag{5.32}$$

for (2n-3) unknowns $\alpha_1 \ldots \alpha_{2n-3}$ with the stipulation $\lambda_1 > 0$. α, β are 2(n-1) dimensional vectors. Thus the procedure is to find the linearized system equation at the u.e.p., find the positive eigenvalue λ_1 and the coefficients $\alpha_1 \ldots \alpha_{2n-3}$. The hyperplane is then given by

$$x^T y_1 = 0 \qquad \text{with} \qquad y_1 = \begin{bmatrix} \alpha_1 \\ \cdot \\ \cdot \\ \cdot \\ \alpha_{2n-3} \\ 1 \end{bmatrix} = \begin{bmatrix} \alpha \\ --- \\ \beta \end{bmatrix} \tag{5.33}$$

$$\text{i.e., } \Delta\delta^T\alpha + \Delta\omega^T\beta = 0$$

A trajectory starting in the hyperplane will stay in the hyperplane and will approach the type-one e.p. as $t \to \infty$. Instead of working with the K matrix, which is of order (2n-2), one can also work with the J matrix whose order is (n-1). Suppose λ_1' is the positive eigenvalue of J, then

$$J^T\beta = \lambda_1' \beta$$

where β is the eigenvector and $\alpha = [\sqrt{\lambda_1' + (C^2/4)} + C/2]\ \beta$. Thus the problem of determining α and β reduces to finding the positive eigenvalue and associated eigenvector of the J^T matrix.

Each tangent hyperplane being an approximation to the stable manifold at an u.e.p. locally describes the stability boundary. Hence, the time when the sustained fault trajectory intersects a tangent hyperplane for the first time gives an estimate of critical clearing time. In Yee and Spalding [1977], they have applied the method to a 15 machine 35 bus system. They identified a total of 10 type-one u.e.p.'s of which nine corresponded to one machine losing synchronism and the tenth corresponded to two machines losing synchronism.

Numerical results reported indicate that this method gives good results for fault at the terminals of the generator since the mode of instability is generally predictable (i.e., the generator adjacent to the fault pulls away from the rest of the system). The results were not that encouraging when the fault was in the middle of the system. This method has been extended by taking second order terms for the hyperplane so that it becomes a hyper surface. The method has been applied to machines with saturable exciters. Also sensitivity of t_{cr} with respect to system parameters and pre-fault loading have also been done. (Yee and Muir [1980], Cook and Eskicioglu [1983])

5.5 Potential Energy Boundary Surface (PEBS) Method

Ever since it was first proposed by Kakimoto et al. [1978] and Athay et al. [1979], the method has received wide attention by researchers because it avoids computing the controlling u.e.p. and needs a quick fault-on system integration to compute V_{cr}. However, except for single machine systems, the method is not very accurate except under certain conditions. Hence, there is continued research on fine-tuning the method. In this section we will first motivate the method through a second order system example, explain under what conditions the method yields reliable results and finally discuss techniques to improve the accuracy using the method.

5.5.1 Motivation via single machine system

If the energy remains constant throughout the motion of a system, the system is said to be conservative. Consider a single machine infinite bus system whose post-fault equations are

$$M \frac{d^2\delta}{dt^2} = P_m - P_e \sin \delta \tag{5.34}$$

The right-hand side of (5.34) can be written as $\frac{-\partial V_{PE}}{\partial \delta}$ where

$$V_{PE} = -P_m \delta - P_e \cos \delta \qquad (5.35)$$

Using (5.35), (5.34) can be rewritten as

$$\frac{d}{dt}\left[\frac{M}{2}\left(\frac{d\delta}{dt}\right)^2 + V_{PE}\right] = 0 \qquad (5.36)$$

It follows from (5.36) that the quantity in the brackets is constant. In (5.34) the equilibrium is $\frac{d\delta}{dt} = 0$ and $\delta = \delta^s = \sin^{-1}\left(\frac{P_m}{P_e}\right)$. If we make a change of coordinates so that $V_{PE} = 0$ at $\delta = \delta^s$, then (5.35) becomes

$$V_{PE} = -P_m (\delta-\delta^s) - P_m (\cos \delta - \cos \delta^s) \qquad (5.37)$$

With this the energy function $V(\delta,\omega)$ in the brackets of (5.36) can be written as

$$V(\delta,\omega) = \frac{1}{2} M\omega^2 - P_m (\delta-\delta^s) - P_{em} (\cos \delta - \cos \delta^s) = E \qquad (5.38)$$

where E is a constant. This is precisely the energy function derived in Chapter 1 via the equal area criterion. E, which is the sum of kinetic and potential energies, remains constant once the fault is cleared. $V(\delta.\omega)$ evaluated at $t = t_{c\ell}$ represents the energy present in the system at $t = t_{c\ell}$. This energy must be absorbed by the system once the fault is cleared if the system is to be stable. The kinetic energy is always positive and is the difference between E and $V_{PE}(\delta)$. This is shown graphically in Fig. 5.11.

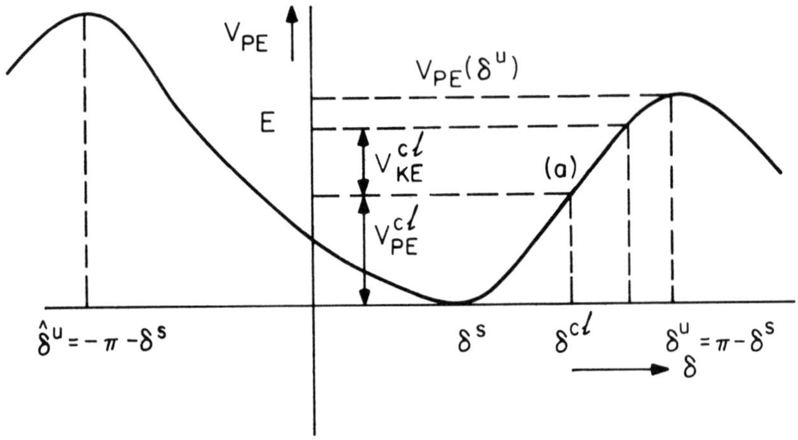

Figure 5.11 Potential energy "well."

At $\delta = \delta^s$ both the KE and PE are zero. Suppose at $t = t_{c\ell}$, corresponding to $\delta = \delta^{c\ell}$, the rotor has a velocity $\omega^{c\ell}$, then

$$V^{c\ell} = \frac{1}{2} M\omega^2_{c\ell} - P\,(\delta^{c\ell} - \delta^s) - P_e\,(\delta^{c\ell} - \delta^s). \tag{5.39}$$

$$= V^{c\ell}_{KE} + V^{c\ell}_{PE}$$

This is the value of E. There are two other equilibrium points of the system at $\delta^u = \pi - \delta^s$ and $\hat{\delta}^u = -\pi-\delta^s$. Both these are unstable and, in fact are type-one (saddle-type) e.p.'s. The potential energy is zero at $\delta = \delta^s$ and has relative maximum at $\delta = \delta^u$ and $\delta = \hat{\delta}^u$. If at point (a), $\omega^{c\ell}$ is known then E is as shown. If $E > V_{PE}\,(\delta^u)$ then the system is unstable if δ increases due to accelerating power. If δ decreases due to deceleration for $t > 0$, then the system is unstable if $E > V\,(\hat{\delta}^u)$. The points δ^u and $\hat{\delta}^u$ constitute the 0-dimensional PEBS for the single machine system. Some researchers restate the above idea by saying that if the V_{PE} is initialized to zero at $\delta^{c\ell}$, $V_{KE}^{c\ell}$ represents the excess kinetic energy injected into the system. Then stabilty of the system is determined by the ability of the post-fault system to absorb this excess K•E (i.e., the system is stable if $V_{PE}(\delta^u) - V_{PE}(\delta^{c\ell}) > V_{KE}^{c\ell}$). In terms of the equal area criterion of Chapter I this means area 2 > area 1.

Most of the stability concepts can be interpreted as if the moment of inertia M is assumed as a particle that slides without friction on a physical "hill" with the shape $V_{PE}(\delta)$. Motions within a potential "well" are bounded and hence stable. It is interesting to relate the potential "well" to stability of equilibrium points for small disturbances.

$$M \frac{d^2\delta}{dt^2} = -\frac{\partial V_{PE}(\delta)}{\partial \delta} \tag{5.40}$$

We can expand the right-hand side of (5.40) in a Taylor series about $\delta = \delta^*$ i.e., $\delta = \delta^* + \Delta\delta$, and retain only the linear term. Then

$$M \frac{d^2\Delta\delta}{dt^2} = -\left.\frac{\partial^2 V_{PE}(\delta)}{\partial \delta}\right|_{\delta^*} \Delta\delta \tag{5.41}$$

i.e.,

$$M \frac{d^2\Delta\delta}{dt^2} + \left.\frac{\partial^2 V_{PE}(\delta)}{\partial \delta}\right|_{\delta^*} \Delta\delta = 0 \tag{5.42}$$

If $\left.\frac{\partial^2 V_{PE}}{\partial \delta}\right|_{\delta^*} < 0$, the equilibrium is unstable. If $\left.\frac{\partial^2 V_{PE}}{\partial \delta}\right|_{\delta^*} > 0$, then it is an oscillatory system and the oscillations around δ^* are bounded. In the case of Equation (5.34) it can be verified that is δ^s is stable, then both δ^u and $\hat{\delta}^u$ are unstable using this criterion.

The energy function, Lyapunov function and the PEBS all are equivalent in the case of a single machine system. It is in the case of multimachine systems and non-conservative systems that each method gives only approximations to the true stability boundary! In the case of multimachine systems, the second derivative of V_{PE} is the Hessian matrix. Marino and Nicosia [1983] show that if the Hessian is positive definite, the system admits of an energy function which is locally positive definite.

5.5.2 Multi-machine PEBS

Consider the multi-machine system of (n+1) internal nodes where (n+1)th node is taken as the reference bus whose M is very large, i.e., $\dot{\omega}_{n+1} = 0$, $\delta_{n+1} = 0$. Also $\omega_{n+1} = \overline{\omega}$ = constant. The equations of the post-fault system are

$$\dot{\delta}_i = \omega_i \qquad i=1,\ 2,\ \dots\ n$$

$$M_i\dot{\omega}_i = P_i - D_i\omega_i - \sum_{j \neq i}^{n+1} C_{ij} \sin \delta_{ij} \tag{5.43}$$

δ_i and ω_i are the rotor angle and relative velocity of the ith machine with respect to the $(n+1)^{th}$ machine. The energy function for this has been derived as (neglecting transfer conductances)

$$V(\delta,\omega) = \frac{1}{2}\sum_{i=1}^{n} M_i\omega_i^2 - \sum_{i=1}^{n} P_i\ (\delta_i - \delta_i^s)$$

$$- \sum_{i=1}^{n}\ \sum_{j=i+1}^{n+1} C_{ij}\left[\cos \delta_{ij} - \cos \delta_{ij}^s\right] \tag{5.44}$$

$$= V_{KE}(\omega) + V_{PE}(\delta)$$

V_{KE} and V_{PE} are the kinetic energy and potential energies respectively. Since there is an infinite bus in the system, it automatically becomes the COI of the system so that $V(\delta,\omega)$ is also the transient energy function with δ,ω being the COI angles and velocities, respectively.

Equation (5.43) can be put in the form

$$\dot{\delta}_i = \omega_i \qquad i=1, 2, \ldots n$$

$$M_i\dot{\omega}_i = - D_i\omega_i - \frac{\partial V_{PE}(\delta)}{\partial \delta_i} \tag{5.45}$$

With $D_i = 0$, note the similarity of this equation to the single machine case.

It has been shown in Pai [1981] that $V(\delta,\omega)$ is a positive definite function for the system since it can be put in a Luré form satisfying the sector condition around (δ^s, ω^s). Now

$$\dot{V}(\delta,\omega) = \frac{\partial V}{\partial \delta}\dot{\delta} + \frac{\partial V}{\partial \omega}\dot{\omega}$$

$$= -\sum_{i=1}^{n} D_i\omega_i^2 \leq 0 \tag{5.46}$$

Hence, $V(\delta,\omega)$ is also a Lyapunov function for the system (5.45).

All the equilibrium points of (5.45) lie on the subspace $\{(\delta,\omega)|\delta\varepsilon R^n, \omega=0\}$. Therefore, the energy function $V(\delta,\omega)$ at the equilibrium point (δ^s,ω^s) is of the following form: $V(\delta^s,\omega^s) = V_{PE}(\delta^s)$. This motivates the study of stability region in the δ subspace and we consider the associated gradient system (Chiang et al. [1987], Thomas and Thorp [1985])

$$\dot{\delta} = - \frac{\partial V_{PE}(\delta)}{\partial \delta} \tag{5.47}$$

We now claim that the stabilty boundary $\partial A(\delta^s)$ of the gradient system for the s.e.p. δ^s is, in fact, the potential energy boundary surface (PEBS). For the system (5.47)

$$\partial A(\delta^s) = \bigcup_i W^s(\delta_i) \tag{5.48}$$

where δ_i's are the equilibrium points on the stability boundary $\partial A(\delta_s)$. Since the gradient of $V_{PE}(\delta)$ is a vector perpendicular to the level surface $\{\delta : V_{PE}(\delta) = C\}$ in the increasing direction, it follows that the vector field at every regular point of the system (5.47) is perpendicular to the level surface $\{\delta : V_{PE}(\delta) = C\}$. Thus, $\partial A(\delta^s)$ intersects the level surface $V_{PE}(\delta) = C$ orthogonally (Fig. 5.12).

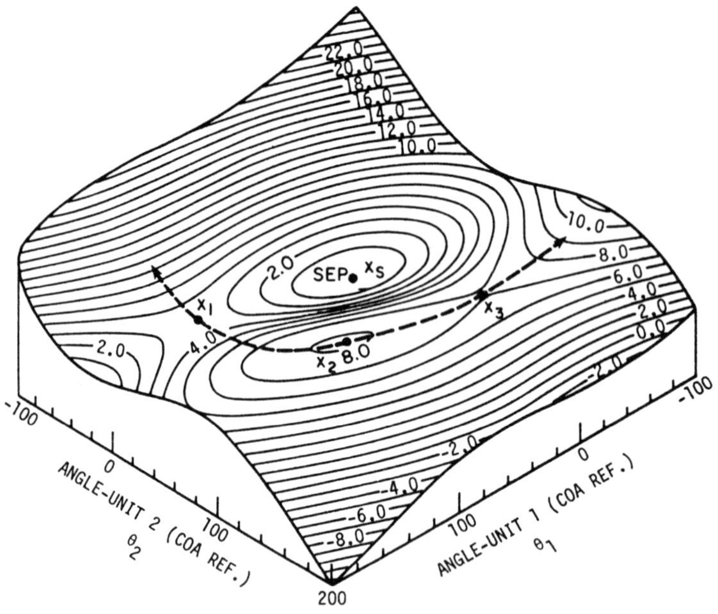

Figure 5.12 Potential energy boundary surface. (From Athay et al. [1979])

Thus if a trajectory crosses the PEBS orthogonally, then V_{PE} achieves the maximum at that point. Kakimoto uses this fact to find when

$$\frac{dV_{PE}}{dt} = \frac{\partial V_{PE}}{\partial \delta} \cdot \frac{d\delta}{dt} = \frac{\partial V_{PE}}{\partial \delta} \cdot \omega$$

is zero. When the trajectory is orthogonal to PEBS, then ω and $\frac{\partial V_{PE}}{\partial \delta}$ are orthogonal (Fig. 5.13). This was the original explanation of Kakimoto [1978].

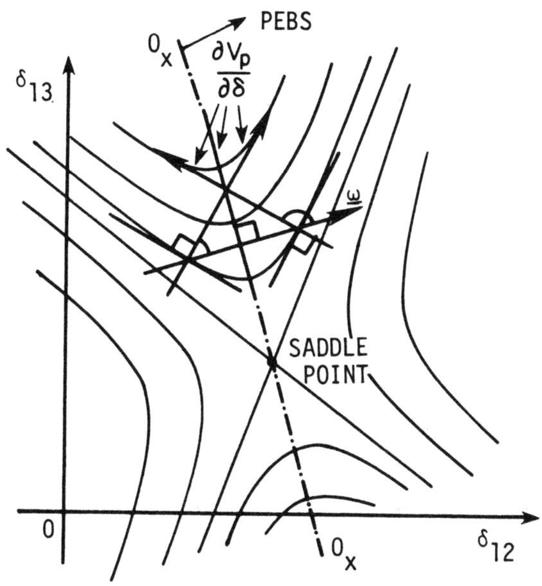

Figure 5.13 Relation between ω and $\partial V_p/\partial \delta$ when the system crosses the PEBS in relative rotor angle space (3 machine system).

If we assume this to be true for the fault-on trajectory also, then at the instant of PEBS crossing t^*, $V_{PE}(\delta^*)$ gives a good estimate of V_{cr}. Another implied assumption is that the critically cleared trajectory reaches the PEBS around the same point and then returns to δ^s. To summarize, Kakimoto's method is

1. Integrate the fault-on trajectory until V_{PE} reaches a maximum in time. This V_{PE}^{max} is an estimate of true V_{cr}.

2. From the fault-on trajectory find when $V(\delta,\omega) = V_{cr}$. This is t_{cr}. This can be shown graphically in Fig. 5.14.

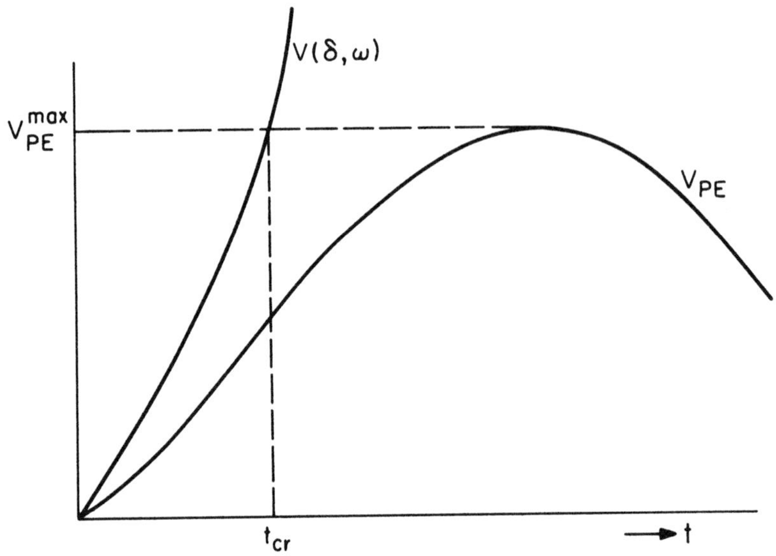

Figure 5.14 - PEBS Method

Athay et al. [1979] indicate a different method of finding when the PEBS is reached. Their method includes transfer conductance terms in Equation (5.43) as well as in the energy function (5.44).

The basic swing equations are

$$M_i \frac{d^2\delta_i}{dt^2} + D_i \frac{d\delta_i}{dt} = P_i - \sum_{j\neq i}^{n} \left(C_{ij} \sin \delta_{ij} + D_{ij} \sin \delta_{ij}\right) \qquad (5.49)$$

$$i=1,\ 2,\ \ldots\ n$$

The COI notation is used where

$$\theta_i = \delta_i - \delta_o , \qquad \delta_o = \frac{1}{M_T} \sum_{i=1}^{n} M_i \delta_i$$

$$\tilde{\omega}_i = \omega_i - \omega_o \qquad \omega_o = \frac{1}{M_T} \sum_{i=1}^{n} M_i \omega_i \tag{5.50}$$

and $M_T = \sum_{i=1}^{n} M_i$

Equation (5.49) is put in the state space notation as

$$\dot{\theta}_i = \omega_i$$

$$M_i \dot{\tilde{\omega}}_i = P_i - P_{ei} - \frac{M_i}{M_T} P_{COI} \overset{\Delta}{=} f_i(\theta)$$

$$i=1,\ 2,\ \ldots\ n \tag{5.51}$$

where $$P_{COI} = \sum_{i=1}^{n} P_i - 2 \sum_{i=1}^{n} \sum_{j=i+1}^{n} D_{ij} \cos \theta_{ij} \tag{5.52}$$

The energy function has been derived in Chapter II as

$$V(\theta,\tilde{\omega}) = \frac{1}{2} \sum_{i=1}^{n} M_i \tilde{\omega}_i^2 - \sum_{i=1}^{n} P_i \,(\theta_i - \theta_i^s)$$

$$- \sum_{i=1}^{n} \sum_{j=i+1}^{n} [C_{ij} \,(\cos \theta_{ij} - \cos \theta_{ij}^s)$$

$$- \int_{\theta_i^s + \theta_j^s}^{\theta_i + \theta_j} D_{ij} \cos \theta_{ij} \, d(\theta_i + \theta_j)] \tag{5.53}$$

The last term on the right-hand side of (5.53) is a path dependent term. Equation (5.53) is re-written as

$$V(\theta,\tilde{\omega}) = V_{KE}(\tilde{\omega}) + V_p(\theta) + V_d(\theta) \tag{5.54}$$

where
$$V_{PE}(\theta) = V_p(\theta) + V_d(\theta). \tag{5.55}$$

$V_d(\theta)$ is the path dependent term and $V_p(\theta)$ is the closed form term in the potential energy function. To construct the PEBS in the θ space linear path angle dependence is assumed in different directions from θ^s. Thus

$$\theta = (\theta - \theta^s)\,\alpha + \theta^s = \Delta\theta\alpha + \theta^s \tag{5.56}$$

Thus
$$-\frac{\partial V_{PE}(\theta)}{\partial \theta_i} = f_i(\theta) \tag{5.57}$$

$$V_{PE}(\theta) = -\int_{\theta_i^s}^{\theta_i} \sum_{i=1}^{n} f_i(\theta)\, d\theta_i \tag{5.58}$$

Assuming linear path dependence

$$V_{PE}^{appr}(\theta) = -\sum_{i=1}^{n} \int_{o}^{1} f_i\,(\Delta\theta\alpha + \theta^s)\,\Delta\theta_i\, d\alpha \tag{5.59}$$

Now $\frac{d}{d\theta} V_{PE}^{appr}(\theta) = 0$ implies

$$-\sum_{i=1}^{n} f_i\,(\Delta\theta\alpha + \theta^s)\,\Delta\theta_i = 0 \tag{5.60}$$

If (5.60) is satisfied for some α, i.e., some θ^*, then it means that θ^* satisfies

$$\sum_{i=1}^{n} f_i\,(\theta)\,(\theta_i - \theta_i^s) = 0 \tag{5.61}$$

This is the characterization of the PEBS. In vector form (5.61) can be written as $f^T(\theta) \cdot (\theta - \theta^s) = 0$. Denoting $\theta - \theta^s = \hat{\theta}$, we shall show by analogy to the zero transfer conductance case that inside the PEBS $f^T(\theta) \cdot \hat{\theta} < 0$ and outside the PEBS it is > 0. In the absence of transfer conductances, $f(\theta) = -\frac{\partial V_{PE}(\theta)}{\partial \theta}$. When θ is away from θ^s within the potential multidimensional "well", $\frac{\partial V_{PE}(\theta)}{\partial \theta}$ (which is the gradient of the PE function) and $\hat{\theta}$ (i.e., $(\theta - \theta^s)$) are both > 0. Hence, $f^T(\theta) \cdot \hat{\theta} < 0$ inside the "well." Outside the "well" $\frac{\partial V_{PE}(\theta)}{\partial \theta} < 0$ and, hence, $f^T(\theta) \cdot \hat{\theta} > 0$. In actual computation $V_d(\theta)$ is computed using the trapezoidal rule. When the fault-on trajectory is known (Athay et al. [1979])

$$V_d(\theta) = \sum_{i=1}^{n-1} \sum_{j=i+1}^{n} I_{ij}$$

where at the mth step

$$\begin{aligned} I_{ij}(m) &= I_{ij}(m-1) \\ &+ \frac{1}{2} D_{ij} \left[\cos\left(\theta_i(m) - \theta_j(m)\right)\right. \\ &\qquad + \cos\left(\theta_i(m-1) - \theta_j(m-1)\right)\big]\big[\theta_i(m) + \theta_j(m) \\ &\qquad\qquad - \theta_i(m-1) - \theta_j(m-1)\big] \end{aligned}$$

where (m) and (m-1) indicate values at m^{th} and $(m-1)^{th}$ steps, respectively. $I_{ij}(0) = 0$

The following are the steps to compute t_{cr} using the PEBS method.

1. Compute the fault-on trajectory

2. Compute $f^T(\theta)\cdot\hat{\theta}$ and $V_{PE}(\theta)$ at each time step. The parameters in $f(\theta)$ and $V_{PE}(\theta)$ pertain to the post-fault configuration.

3. Inside the potential "well" $f^T(\theta)\cdot\hat{\theta} < 0$. Continue steps 1 and 2 until $f^T(\theta)\cdot\hat{\theta} = 0$. This is the PEBS crossing $(t^*, \theta^*, \tilde{\omega}^*)$. At this point find $V_{PE}(\theta^*)$. This is a good estimate of V_{cr} for that fault.

4. Find when $V(\tilde{\theta}, \tilde{\omega}) = V_{cr}$ on the fault-on trajectory. This gives t_{cr}.

The intuition behind using V_{PE}^{max} along the faulted direction is analogous to that in one machine case, namely that V_{PE}^{max} represents the maximum system energy at $t = t_{c\ell}$, that can be converted into potential energy before the system becomes unstable.

5.6 Hybrid Method Using the Gradient System

A third interpretation and modification of the PEBS method is due to Chiang et al. [1988]. They, first of all, make the following observations.

1. The unstable equilibrium points of the gradient system (5.47) are the same as those of the original dynamical system (5.45) with $\omega = 0$.

2. The stability boundary of the original system is $\partial A(\delta^s,0) = \bigcup_{\substack{i \\ (\delta_i,0)\,\in\,\partial A(\delta^s,0)}} W^s(\delta_i,0)$

   This is the same result as in Sec. 5.2 of this chapter.

3. The stability boundary $\partial A_g(\delta^s)$ of the gradient system (5.47) given by

$$\partial A_g(\delta^s) = \bigcup_i W^s(\delta_i)$$

   is the PEBS.

Conceptually this is also the PEBS definition due to Kakimoto et al. [1978] and Athay et al. [1979]. Note that the PEBS is not the intersection of the stability boundary $\partial A(\delta^s,0)$ with the subspace $[(\delta,\omega):\omega=0]$.

We will explain now the conditions under which PEBS method will give good results. It is a graphical illustration. For mathematical proof, etc., see Chiang et al. [1988].

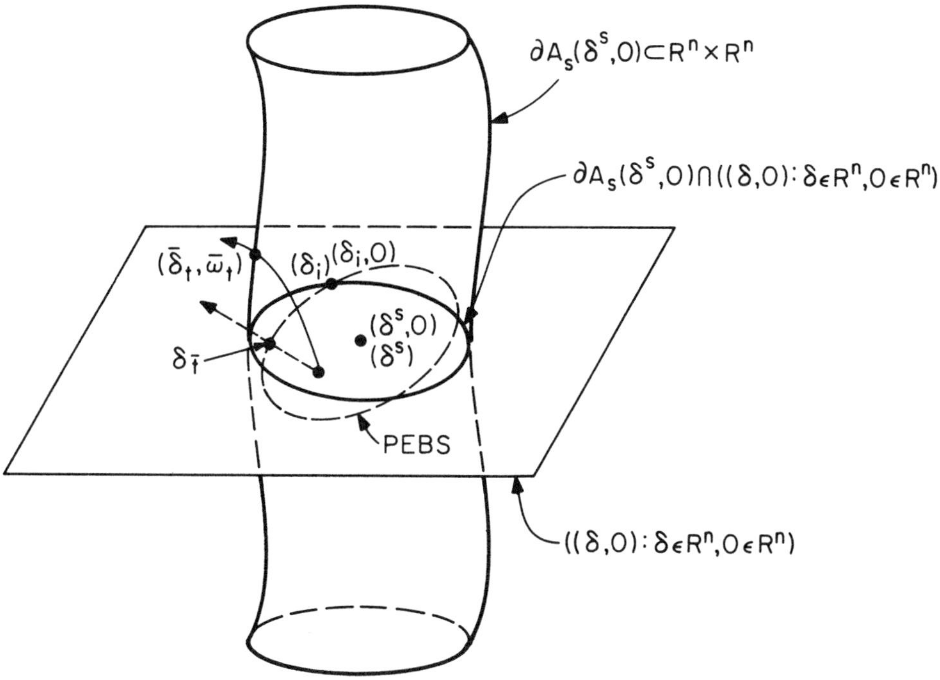

Figure 5.15 The fault-on trajectory $(\delta(t),\omega(t))$ intersects the stability boundary $\partial A_s(\delta_s,o)$ of the original system at $(\overline{\delta_t},\overline{\omega_t})$. The projection of the fault-on trajectory on the δ-space intersects the PEBS at $(\delta_{\overline{t}})$. (From Chiang et al. [1988])

Given a fault-on trajectory $(\delta(t), \omega(t))$, let $(\bar{\delta}(t), \bar{\omega}(t))$ denote the exit point on the stability boundary of the original system and let $\delta_{\bar{t}}$ denote the exit point of the projected trajectory $\delta(t)$ in δ space of the gradient system (Fig. 5.15). Let $\partial V(x)$ denote the connected constant energy surface $V(\cdot)$ containing the stable equilibrium point $(\delta^s, 0)$ passing through the point x. The idea of the PEBS is to use $\partial V(\delta_{\bar{t}}, 0)$ to approximate the relevant stability boundary of the original system. Under what conditions is this approximation good? In Chiang et al. [1988] it is shown that the approximation is good if the following two conditions are satisfied. Let $(\hat{\delta}, o)$ be the controlling u.e.p.

(i) $(\delta_{\bar{t}})$ is on the stable manifold of $(\hat{\delta})$ of the gradient system implies $(\bar{\delta}_t, \bar{\omega}_t)$ is on the stable manifold of $(\hat{\delta}, 0)$ of the original system.

(ii) The fault-on trajectory $(\delta(t), \omega(t))$ passes through the constant energy surface $\partial V(\delta_{\bar{t}}, 0)$ before it passes through the stable manifold $W^s(\hat{\delta}, 0)$.

Notice that it is difficult to check condition (ii) since it implies knowing the stable manifold of $W^s(\hat{\delta}, 0)$!

We now state the original PEBS method due to Kakimoto et al. [1978] and the modified PEBS method due to Chiang et al. [1988].

Original PEBS Method

Step 1 From the direction of the fault-on trajectory $(\delta(t), \omega(t))$, detect the point δ^* at which the projected $\delta(t)$ crosses PEBS. We have indicated earlier two methods to detect the PEBS crossing, namely,

(i) when V_{PE} is maximum and

(ii) when $f^T\hat{\theta}$ changes sign from - to +, both along the fault-on trajectory. Let the value of $V_p(\cdot)$ at δ^* be $V_{cr} = V_p(\delta^*)$.

Step 2 Use the connected constant energy surface of the set $\{(\delta,\omega): V(\delta,\omega) = V_p(\delta^*)\}$ containing the stable equilibrium point as the local approximation of the stability boundary.

Modified PEBS Method

Step 1' From the fault-on trajectory detect the point δ^* at which the projected trajectory crosses PEBS. Let the value of $V_p(\cdot)$ at δ^* be $V_{cr} = V_p(\delta^*)$. This is the same as Step 1 of the original PEBS method.

Step 2' Find the equilibrium point of the gradient system whose stable manifold contains δ^*, say $\hat{\delta}$. Let the value of $V_p(\cdot)$ at $\hat{\delta}$ be $V_{cr} = V_p(\hat{\delta})$.

Step 3' Use the connected constant energy surface of the set $\{(\delta,\omega): V(\delta,\omega) = V_p(\hat{\delta})\}$ containing the stable equilibrium point as the local approximation of the stability boundary $\partial A(\delta^s,0)$.

Comments

1. To implement Step 2', we can use δ^* as the initial condition and integrate the post-fault gradient system until the local minimum of $\left|\frac{\partial V_p(\delta)}{\partial \delta}\right|$ is achieved, say at δ_0. Then use δ_0 as the initial guess to solve the algebraic equations of the post-fault gradient system

$$\frac{\partial V_p(\delta)}{\partial \delta} = 0$$

to obtain $\hat{\delta}$.

2. $\hat{\delta}$ in the above procedure is the same as the controlling u.e.p. δ^* in many cases may be quite close to $\hat{\delta}$ so that a post-fault load flow solution of the gradient system using δ^* as the initial guess may be adequate.

Conclusion

This chapter has focused mainly on characterizing the region of stability of a stable equilibrium point of the classical model of a multi-machine system. Both the closest u.e.p. and controlling u.e.p. method were discussed. Next we discussed the PEBS approach and its modification for the computation of t_{cr}. It turns out that the modificaton of the PEBS method ends in computing the controlling u.e.p. It is recognized, however, that the critically cleared trajectory exits the stability-boundary at an exit point $(\overline{\delta}(t), \overline{\omega}(t))$ whose computation is possible only if we have an expression for the stable manifold to which the exit point belongs. In this context the method of hyperplanes and hypersurfaces was also discussed.

Chapter 6

PRACTICAL APPLICATIONS OF THE ENERGY FUNCTION METHOD

6.1 Introduction

In the previous chapters we have discussed the theoretical aspects of energy function analysis of power systems. These were basically (i) motivation for energy function analysis through equal area criterion, (ii) internal node and structure preserving energy functions, (iii) reduced order energy functions, (iv) energy functions for detailed models, and, finally, (v) the characterization of the stability region of the post-fault stable equilibrium point and justification of the PEBS method. It is fair to say that in the power system stability area theory has sometimes tended to be ahead of practical applications. The final aim of any theoretical advance is the possibility for application for real world systems.

The last decade has been an active period for applications of the Lyapunov/ energy function for practical and large scale power systems. In this chapter we discuss application related issues. These are:

(i) Application of the controlling u.e.p., the PEBS method and the acceleration method to practical large scale systems in computing t_{cr}, energy margins etc.

(ii) Use of energy function method in dynamic security assessment.

This chapter is written in a manner so that one can read this chapter going through Chapter I and the initial portion of Chapter II pertaining to internal node energy functions. Hence there is some overlap with Chapter V.

6.2 The Controlling u.e.p. Method

We first discuss the controlling u.e.p. method algorithm due to Athay et al. [1979].

The energy function in the COI notation for a n-machine system in COI notation (Eq. (2.24)) is

$$V(\theta,\omega) = V_{KE}(\tilde{\omega}) + V_{PE}(\theta) \tag{6.1}$$

where
$$V_{PE}(\theta) = V_p(\theta) + V_d(\theta) \tag{6.2}$$

and
$$V_{KE}(\tilde{\omega}) = \frac{1}{2}\sum_{i=1}^{n} M_i\tilde{\omega}_i^2 \tag{6.3}$$

$$V_p(\theta) = -\sum_{i=1}^{n} P_i(\theta_i - \theta_i^s) - \sum_{i=1}^{n}\sum_{j=i+1}^{n} C_{ij}(\cos\theta_{ij} - \cos\theta_{ij}^s) \tag{6.4}$$

$$V_d(\theta) = \int_{\theta_i^s+\theta_j^s}^{\theta_i+\theta_j} D_{ij}\cos(\theta_{ij})\, d(\theta_i + \theta_j) \tag{6.5}$$

The C_{ij} and D_{ij} parameters are for the post-fault configuration.

The method consists of the following steps.

1. For a given contingency which involves either line switching or load/generation change, we need to compute the post disturbance s.e.p. θ^s.

2. Next compute the controlling u.e.p. θ^u.

3. V_{cr} is approximated as $V_{cr} = V(\theta^u, 0)$.

4. The fault-on trajectory in the case of a fault or the post disturbance trajectory in the case of load/generation change is computed and the values of θ, $\tilde{\omega}$ so obtained are substituted in the V-function until $V(\theta, \tilde{\omega}) = V_{cr}$. In the case of a fault this time instant gives t_{cr} and the system is stable if the fault is set to clear at a time $< t_{cr}$. In the case of load/generation change, the system is stable if $V(\theta, \tilde{\omega}) < V_{cr}$ for all t.

Most of the above steps are straightforward except Step 2. Surrounding θ^s there are a number of u.e.p.'s. The faulted trajectory will cross the stable manifold of a controlling u.e.p. as explained in the previous chapter. Since there is no a-priori way to know which among the u.e.p.'s is the controlling one, several methods have been proposed. Both the s.e.p. and u.e.p.'s are solutions of the real power equations,

$$f_i(\theta) = P_i - P_{ei}(\theta) - \frac{M_i}{M_T} P_{COI}(\theta) = 0 \qquad i=1, 2, \ldots n \qquad (6.6)$$

Since $\theta_n = \frac{-1}{M_n} \sum_{i=1}^{n-1} M_i \theta_i$, it is sufficient to solve for

$$f_i(\theta) = 0 \qquad i=1, 2, \ldots n-1 \qquad (6.7)$$

with θ_n being substituted in the equations (6.7) in terms of $\theta_1 \ldots \theta_{n-1}$.

Computation of s.e.p.

Generally the s.e.p. θ^s is close to θ^o the prefault e.p. Hence, using θ^o as the starting point, Eq. (6.7) can be solved using the Newton-Raphson method. At any equilibrium point θ^e

$$P_i - P_{ei}(\theta^e) - \frac{M_i}{M_T} P_{COI}(\theta^e) = 0 \tag{6.8}$$

Since by assumption in the classical model, P_{mi}'s are constant, we have

$$\frac{P_i - P_{ei}(\theta^e)}{P_j - P_{ej}(\theta^e)} = \frac{M_i}{M_j} \tag{6.9}$$

Once θ^s is computed, the injected real powers at the internal nodes are calculated in the COI reference frame as

$$P_{gi}(\theta^s) = \text{Re}\left[\,|E_i|\,\angle\theta_i^s\; I_i^*\right]$$

$$= \text{Re}\left[\,|E_i|\,\angle\theta_i^s \sum_{j=1}^{n} Y_{ij}^* E_j^*\right]$$

$$\triangleq P_{ei}(\theta^s) + E_i^2 G_{ii} \tag{6.10}$$

Also $$\frac{P_i - P_{ei}(\theta^s)}{M_i} = \frac{P_{COI}(\theta^s)}{M_T} \tag{6.11}$$

Equation (6.11) shows that all the absolute accelerations are zero since $P_{ei}(\theta^s) = P_{ei}(\delta^s)$. This follows from the fact that P_{ei} is a function of the differences of δ_i's or equivalently θ_i's. Hence, the system is in synchronous equilibrium. At the stable operating point $P_{COI}(\theta^s) = 0$ and there is no net accelerating power on the COI. Thus the system is also in frequency equilibrium.

To find the controlling u.e.p., the following strategy is employed.

1. Integrate the fault-on system equations using regular integration technique or a trajectory approximation. This could be either a cosine approximation or simply a Taylor series expansion of θ, $\tilde{\omega}$ around $t = 0$.

2. At each step of the integration, compute $F(\theta) = \sum_{i=1}^{n} f_i^2(\theta)$ which is the power mismatch of the post-fault system. When $F(\theta)$ reaches a maximum, this implies we have reached or come close to the boundary of the principle singular surface. The concept of principle singular surface for multi-machine systems is quite complex. In the case of a single machine it is the point of steady state instability, i.e., $\delta = \pi/2$. This can also be inferred from the case of multimachine systems if $D_{ij} \equiv 0$. In this instance the gradient of $V_p(\theta)$ is the negative of the vector of accelerating powers $f_i(\theta)$. Hence, $F(\theta)$ can be interpreted as the Euclidean norm squared of the gradient of the potential energy function. Beyond the singular surface, $F(\theta)$ starts decreasing as can be inferred from the single machine analogy. Hence, on the boundary of the principle surface, $F(\theta)$ is maximum.

3. The value of θ where $F(\theta)$ is maximum is denoted by θ^{ss}. Starting with θ^{ss}, a vector $\theta^{ss} - \theta^{s}$ is formed and is normalized to form the direction vector h, i.e., $h = \dfrac{\theta^{ss} - \theta^{s}}{|\theta^{ss} - \theta^{s}|}$.

4. Solve the one-dimensional minimization problem,

$$\underset{z > 0}{\text{Min}}\ F(\theta(z)) \triangleq F(\theta(z^*)) \tag{6.12}$$

where $\theta(z) \triangleq \theta^{ss} + z \cdot h$ and $\theta(z^*) \triangleq \hat{\theta}^u$. With $\hat{\theta}^u$ as the starting point, solve the real power flow equations (6.7) using the Davidon Fletcher Powell (DFP) minimization technique to obtain the controlling u.e.p. θ^u by minimizing $F(\theta)$. The Newton-Raphson method can be used instead of the DFP method but it is known to cause difficulty in convergence. $V(0,\theta^u)$ is an estimate of V_{cr}.

Comments and Remarks

The point where $F(\theta)$ is maximum signifies that maximum synchronizing capability of the post-fault system is reached. Also by this time, the direction in which the faulted trajectory is heading will have been established. The direction vector, h, is one possible choice among several for converging to the controlling u.e.p. $V_{PE}(\theta)$ is path dependent because of $V_d(\theta)$ and since we are not computing any trajectory beyond θ^{ss}, we use the following approximation for $V_d(\theta^u)$ to compute V_{cr}.

$$V_{cr} = V_p(\theta^u) + V_d(\theta^u) \tag{6.13}$$

where
$$V_d(\theta^u) = \sum_{i=1}^{n-1} \sum_{j=i+1}^{n} \frac{\theta_i^u + \theta_j^u - \theta_i^s - \theta_j^s}{\theta_i^u - \theta_j^u - \theta_i^s + \theta_j^s} \left[\sin\theta_{ij}^u - \sin\theta_{ij}^s\right] D_{ij} \tag{6.14}$$

6.3 Modifications to the Controlling u.e.p. Method

Three modifications to the controlling u.e.p. method have been proposed in recent years by Fouad et al. [1987], Vittal et al. [1988] and Chiang, Wu and Varaiya [1988]. The third one combined it with the PEBS technique and was explained in the previous chapter. The first two methods are almost the same except that the former uses possible "candidate" u.e.p.'s as input to the program and lets the program choose the correct u.e.p. while the latter uses an algorithm to sort out the machines into two groups automatically.

6.3.1. Method of Fouad et al. [1987]

The energy function is the same as in the previous section. The emphasis is on judging if the system is stable or not for a given t_{cl} and then computing the energy margin. (For details see EPRI report by Carvalho et al. [1986])

1. For a given fault a set of possible candidate u.e.p.'s are chosen and these constitute an input to the Transient Energy Function (TEF) program.

   Let θ^s be the post fault s.e.p. This is computed as explained in the previous section. For illustration purposes let us limit to three possible modes of instability (MOI) in a 20 machine system. These are

   (i) Machines 1, 3, 5 going unstable

   (ii) Machines 5, 10, 16 going unstable

   (iii) Machines 5, 12, 18 going unstable

   Then the "candidate" approximate u.e.p.'s $\hat{\theta}^u_{(1)}$, $\hat{\theta}^u_{(2)}$ and $\hat{\theta}^u_{(3)}$ are

   (i) $(\pi - \theta_1^s), \theta_2^s \ldots, (\pi - \theta_3^s), \theta_4^s, (\pi - \theta_5^s) \ldots\ldots \theta_{20}^s$

   (ii) $\theta_1^s \ldots \theta_2^s \ldots\ldots (\pi - \theta_5^s) \ldots\ldots (\pi - \theta_{10}^s) \ldots (\pi - \theta_{16}^s) \ldots\ldots \theta_{20}^s$

   (iii) $\theta_1^s \ldots\ldots (\pi - \theta_5^s) \ldots\ldots (\pi - \theta_{12}^s) \ldots (\pi - \theta_{18}^s) \ldots \theta_{20}^s$.

Although these are referred to as corner points in Fouad et al. [1987], they are not the same as the corner points originally introduced by M. Pavella [1976] and Pai [1981]. Hence we shall refer to them as pseudo-corner points in the sense that if one considers each machine against an infinite bus in turn then $(\pi-\theta_i^s)$ is an unstable equilibrium point. The true corner points are the points where the sine nonlinearity violates the sector condition if the system equations are cast in the Luré-Popov form with no transfer conductances. Let the chosen candidate u.e.p.'s be denoted by their approximate pseudo-corner point as $\hat{\theta}^u_{(k)}$ (k=1, 2, ... ℓ) where ℓ = number of "candidate" u.e.p's chosen by the user, based on past experience.

2. For a given fault using constant acceleration, the faulted trajectory is integrated up to $t_{c\ell}$ and $\tilde{\omega}_{c\ell}$ and $\theta_{c\ell}$ are obtained.

3. Each "candidate" u.e.p. $\hat{\theta}^u_{(k)}$ is now corrected to account for the motion from θ^s to $\hat{\theta}^u_{(k)}$. This is done as follows. Let Group I denote the critical machines (or advanced machines) which can belong to various plants in the system. Group II are the rest of the machines. Let $M_I = \sum_{i\varepsilon I} M_i$ and $M_{II} = \sum_{i\varepsilon II} M_i$.

   (a) At θ^s compute the COI's of the two groups as

$$\theta_I^s = \frac{1}{M_I} \sum_{i\varepsilon I} M_i \theta_i^s \tag{6.15}$$

$$\theta_{II}^s = \frac{1}{M_{II}} \sum_{i\varepsilon II} M_i \theta_i^s \tag{6.17}$$

$$\theta_{I-II}^{s} = \theta_{I}^{s} - \theta_{II}^{s} \tag{6.17}$$

Compute $$\Delta\theta_{I} = (\pi - 2\,\theta_{I-II}^{s}) \frac{M_{II}}{M_{I} + M_{II}} \tag{6.18}$$

and $$\Delta\theta_{II} = (\pi - 2\,\theta_{I-II}^{s}) \frac{M_{I}}{M_{I} + M_{II}} \tag{6.19}$$

(b) Let the COI's of the two groups at the pseudo-corner points be $\hat{\theta}_{I}^{u}$ and $\hat{\theta}_{II}^{u}$.

$$\sum_{i=1}^{n} M_i\,\hat{\theta}_i^{u} = 0 \tag{6.20}$$

i.e. $$M_{I}\,\hat{\theta}_{I}^{u} + M_{II}\,\hat{\theta}_{II}^{u} = 0$$

(c) Calculate the corrected approximate u.e.p. The pseudo-corner point corrected for the motion of the inertial center is given by

$$\hat{\theta}_i^{u} = \theta_i^{s} + \Delta\theta_{I} \qquad i \in I \tag{6.21}$$

$$\hat{\theta}_i^{u} = \theta_i^{s} + \Delta\theta_{II} \qquad i \in II \tag{6.22}$$

Equations (6.21) and (6.22) can be verified by assuming one machine in the critical group and letting $M_{II} \to \infty$. We then have a SMIB system and $\hat{\theta}_1^{u} = (\pi - \theta_1^{s})$ which is correct. Thus the pseudo-corner point is corrected for the motion of the COI to satisfy (6.20)

4. In the angular space θ, maximize V_{PE} along the direction of $\hat{\theta}_{(k)}^{u} - \theta^{s}$ where $\hat{\theta}_{(k)}^{u}$ is now the corrected candidate u.e.p. This is a one-dimensional maximization of

$$V_{PE,Ray}(\alpha) = -\sum_{i=1}^{n} \int_{0}^{\alpha} f_i\,(\theta^{s} + \Delta\theta . p)\,\Delta\theta_i . dp \tag{6.23}$$

and $$\frac{dV_{PE,Ray}}{d\alpha} = -\sum_{i=1}^{n} f_i\,(\theta^{s} + \Delta\theta \cdot \alpha) \cdot \Delta\theta_i$$

It is conjectured in Fouad et al. [1987] that the point θ_{Ray} found in the direction of $\hat{\theta}^u_{(k)} - \theta^s$ where V_{PE} reaches a maximum lies on the PEBS and that it offers a more realistic starting point for computing the controlling u.e.p. We thus get $\hat{\theta}_{Ray,k}$ for each "candidate" u.e.p.

5. Now the selection of the controlling u.e.p. is made based on using the corrected "candidate" u.e.p. $\hat{\theta}_{Ray,k}(k=1,2\ldots\ell)$. The normalized potential energy margin is computed at each corrected candidate u.e.p. and the corrected candidate u.e.p. with the lowest potential energy margin is picked as the starting point for computing the controlling u.e.p. The normalized potential energy in the case of a single machine infinite bus system defined as (Fig. 1.4)

$$\frac{A_2}{A_1} = \frac{V^u_{PE} - V^{c\ell}_{PE}}{V^{c\ell}_{KE}} \triangleq \Delta V_{PE}\big|_n \tag{6.24}$$

In the case of a multimachine system, it is derived as follows:

$$\begin{aligned}\Delta V_{PE} &= V_{PE}{}^u - V_{PE}{}^{c\ell} \\ &= -\sum_{i=1}^{n} P_i(\hat{\theta}_i{}^u - \theta_i{}^{c\ell}) - \sum_{i=1}^{n-1}\sum_{j=i+1}^{n}\{C_{ij}\,(\cos\hat{\theta}_{ij}{}^u - \cos\theta_{ij}{}^{c\ell}) \\ &\quad - D_{ij}\frac{(\hat{\theta}_i{}^u + \hat{\theta}_j{}^u - \theta_i{}^{c\ell} - \theta_j{}^{c\ell})}{(\hat{\theta}_{ij}{}^u - \theta_{ij}{}^{c\ell})}(\sin\hat{\theta}_{ij}{}^u - \sin\theta_{ij}{}^{c\ell})\}\end{aligned} \tag{6.25}$$

The $\hat{\theta}^u_i$ in the above expressions are the components of the corrected candidate u.e.p. $\hat{\theta}_{Ray}$. Based on the corrected "candidate" u.e.p. chosen, the machines are grouped as (i) Advanced machines and (ii) Rest of the machines. In step 1, for example, the advanced machines for the three cases are (1, 3, 5), (5, 10, 16) and (5, 12, 18) respectively. From this two group cluster the equivalent K•E is calculated based on the respective COI's. From Eq. (3.10) we get

$$V_{KE}\Big|_{corr} = \frac{1}{2}\,\frac{M_{cr}\;M_{sys}}{M_{cr}+M_{sys}}\left(\tilde{\omega}^{c\ell}_{cr} - \tilde{\omega}^{c\ell}_{sys}\right)^2 \qquad (6.26)$$

Incidentally, this is also the slow K•E for the two machine grouping. Thus for each corrected candidate u.e.p., $\theta_{Ray,k}$ (k=1, ... ℓ) we compute

$$\Delta V^{k}_{PE}\Big|_{n} = \frac{\Delta V^{k}_{PE}}{V^{k}_{KE}\Big|_{corr}} \quad (k=1,\ 2\ \ldots\ \ell) \qquad (6.27)$$

Pick the $\hat{\theta}_{kay,k}$ that has the lowest normalized $\Delta V_{PE}\big|_n$. Physically it appears that this corresponds to the weakest cutset although this point needs more investigation. This $\hat{\theta}_{Ray,k}$ is now used as a starting point to compute the controlling u.e.p. The nonlinear algebraic equations are

$$P_i - P_{ei} - \frac{M_i}{M_T} P_{COI} = 0\ ,\ i=1,\ 2 \ldots\ n-1 \qquad (6.28)$$

where $\theta_n = -\frac{1}{M_n}\sum_{i=1}^{n} M_i\,\theta_i$

Three techniques have been tried

1) Newton-Raphson method
2) Davidon-Fletcher-Powell (DFP) Minimization method
3) Corrected Gauss-Newton (CGN) method.

Technique (1) was found to give rise to difficulty in convergence in Athay et al. [1979]. Hence, they used the technique (2). But Fouad et al. [1987] report that technique (2) does not work well with large systems. They had good success with the corrected Gauss-Newton method (Gill and Murray [1978]).

6. After θ^u is computed, the transient energy margin is computed as $\Delta V = V(0,\theta^u) - V(\tilde{\omega}^{c\ell}, \theta^{c\ell})$. If $\Delta V > 0$, the system is stable and unstable if $\Delta V < 0$. In order to rank disturbances at different locations in terms of severity one should normalize it by dividing ΔV by $V_{KE}|_{corr}$, i.e., the normalized energy margin is $\Delta V_n = \frac{\Delta V}{V_{KE}|_{corr}}$

The preceding algorithm, although it looks complicated, is based on a lot of physical insight into the behavior of the system and supported by extensive testing. Table 6.1 shows the CPU times for different sized systems.

6.3.2 Method of Vittal et al [1988]

The method of the previous section requires the candidate u.e.p's to be supplied by the analyst. However in a large stressed system when the disturbance may result in the separation of a large group of machines (e.g. inter area mode), not possible. A scheme for automatically generating the correct u.e.p for a given disturbance depends on identifying the generator groupings most seriously affected by the disturbance and makes use of the following information at $t_{c\ell}$.

(i) Kinetic energy of the machines.

(ii) Acceleration of the machines. Once this grouping is done, the rest of the procedure is same as in Sec. 6.3.1. For details of sorting the advanced machines automatically refer to Vittal et al [1988].

TABLE 6.1

CPU times using TEF method.
(Reproduced from Fouad et al. [1987])

VAX/8600 CPU TIME FOR VARIOUS TASKS (IN SECONDS)

Test System Task	17-gens. 40-gens. 117-lines	50-gens. 145-buses 649-lines	100-gens. 1095-buses 2041-lines	150-gens. 1145-buses 2535-lines	228-gens. 1644-buses 3666-lines
1. **TEF Analysis**					
LF Input Data & Network Reduction	1.2	3.0	25	138	281
SEP Solution	0.1	1.0	5	33	82
UEP Solution	0.1	1.5	5	25	62
Other Tasks	1.0	2.0	5	6	7
TOTAL (TEF)	2.4	7.5	40	202	431
2. **Step-by-Step Method**	2.5	14.0	130	171	231

6.4 Potential Energy Boundary Surface (PEBS) Method

This method has been discussed extensively in Chapter V in the context of stability boundary characterization. It has also been discussed in Chapter IV while discussing the generalized potential energy. On large-scale systems the testing of the method is reported by Kakimoto et al. [1984]. Specifically they have tested the method on a 107 Gen, 363 buses and 372 lines and report that it takes 20% of the time taken by one simulation. They used the flux decay model and also exploited the sparsity of the system.

6.5 Mode of Instability (MOI) Method

It is generally assumed that when a system loses stability the system splits up into two groups, the critical generator group and the rest of the system. If this is the case then the generators in each group must be combined to form an equivalent generator. Then we have a two-machine system which in turn can be reduced to a single machine infinite bus (SMIB) system. The idea of such grouping dates back to the original work by Kimbark [1948]. In the context of energy function methods, it was first proposed by Fouad and Vittal [1983] and later on has been pursued by Rahimi and Schaffer [1987], Xue et al. [1988], Lemmon et al. [1988] and Padiyar and Ghosh [1987]. The appeal in using the MOI-based approach is that the analysis of the SMIB system either using the u.e.p. method or the PEBS method is equivalent and straightforward.

For details of specific algorithms for grouping, the reader is advised to refer to the individual references since a number of physical assumptions and heuristic arguments are advanced. We discuss below the theory once the groups are identified. We consider the classical model. We follow the development as in Xue et al. [1988]. Consider a group of critical generators belonging to the subset K and the rest of the generators to the subset S so that we have the model

$$M_k \ddot{\delta}_k = P_{mk} - P_{ek} \qquad k \in K \tag{6.29}$$

$$M_\ell \ddot{\delta}_\ell = P_{m\ell} - P_{e\ell} \qquad \ell \in S \tag{6.30}$$

We define a COI for the subsets K and S in the usual manner i.e.,

$$M_K = \sum_{k \in K} M_k, \quad M_S = \sum_{\ell \in S} M_\ell \tag{6.31}$$

$$\delta_{COI-K} = \frac{1}{M_K} \sum_{k \in K} M_k \delta_k, \quad \delta_{COI-S} = \frac{1}{M_S} \sum_{\ell \in S} M_\ell \delta_\ell \tag{6.32}$$

Adding all the equations in the set K and S separately we get

$$M_K \ddot{\delta}_{COI-K} = \sum_{k \varepsilon K} P_{mk} - \sum_{k \varepsilon K} P_{ek} \tag{6.33}$$

$$M_S \ddot{\delta}_{COI-S} = \sum_{\ell \varepsilon S} P_{m\ell} - \sum_{\ell \varepsilon S} P_{e\ell} \tag{6.34}$$

Let the relative angle between the COI's of the set K and S be

$$\delta = \delta_{COI-K} - \delta_{COI-S} \tag{6.35}$$

Then from (6.33) and (6.34), we get

$$M\ddot{\delta} = P_m - \left[P_c + P_{max} \sin (\delta - \nu)\right] \tag{6.36}$$

where $$M = \frac{M_K M_S}{M_K + M_S} \tag{6.37}$$

$$P_m = \frac{M_S \sum_{k \varepsilon K} P_{mk} - M_K \sum_{\ell \varepsilon S} P_{m\ell}}{M_K + M_S} \tag{6.38}$$

$$P_c = \frac{M_S \sum_{j,k \varepsilon K} E_j E_k G_{jk} - M_K \sum_{j,\ell \varepsilon S} E_j E_\ell G_{j\ell}}{M_K + M_S} \tag{6.39}$$

$$P_{max} = \left(C^2 + D^2\right)^{1/2} \tag{6.40}$$

where $$C = \frac{(M_S - M_K)}{M_S + M_K} \sum_{k \varepsilon K, \ell \varepsilon S} E_k E_\ell G_{k\ell} \tag{6.41}$$

$$D = \sum_{k \varepsilon K} \sum_{\ell \varepsilon S} E_k E_\ell B_{k\ell} \tag{6.42}$$

and $$\nu = - \tan^{-1} \left(\frac{C}{D}\right) \tag{6.43}$$

In the derivation of expressions (6.38)-(6.43) it has been implicitly assumed that generators in each of the two groups are coherent. A general expression in a multi-area framework is given by Eq. (3.40) using slow-coherency concepts. For the specific case of a single machine in the critical group, a general expression is given in Rahimi and Schaffer [1987]. However, for actual calculations, coherency has been assumed in the paper. Otherwise the angle ν and P_C become time varying. By making the coherency assumption we get the SMIB system. For the SMIB system, one can analyze the stability either by equal area criterion or the energy function method discussed in Chapter I. We now discuss briefly the methodology and results based on the work of Xue et al. [1988].

6.5.1 Method of Xue et al. [1988]

This method requires, as in the method of Fouad et al. [1987] (Sec. 6.3.1), that the user supply a list of "candidate" critical clusters, i.e., the set of machines that are likely to separate. As the authors claim the acceleration of machines at $t = 0^+$ can form a basis to develop a rule regarding possible critical clusters. The following observations are made in the paper.

1. For single machine clusters the critical machine is not necessarily close to the fault location.

2. The critical cluster is rather independent of the operating point.

3. When there is more than one machine in a critical cluster, the fault location is at the non-generator buses.

Thus the choice of "candidate" critical clusters must be largely based on experience and to a certain extent on the fault location. An on-line method of choosing the critical cluster is discussed in Vittal et al. [1988] and Lemmon et al. [1988]. These are based on computations of acceleration, K•E of machines etc. at $t = t_{c\ell}$.

Algorithm

1. Assume that "candidate" critical clusters are chosen as input to the program by the user.

2. For each cluster compute the critical clearing time (CCT) using the SMIB equivalent. The method is discussed in Sec. 3.6.

3. The one that gives the lowest critical clearing time is the critical cluster.

The algorithm is illustrated in Fig. 6.1.

Remarks:

1) Note that in Fouad et al. [1987], they use the normalized P•E margin to select the controlling u.e.p. They do not reduce the system to a SMIB system but work with the system energy.

2) Strictly speaking, since ν is time varying, one can still work with the SMIB system if correction to P•E term is made.

3) The algorithm has been tested with variations to detect the second swing instability.

4) The computing times are claimed to be extremely modest. For example the method takes approximately 1/3 of the time taken for the PEBS method which is fast compared to the controlling u.e.p. method.

5) The computed critical clearing times are shown to be extremely accurate compared to the step by step method.

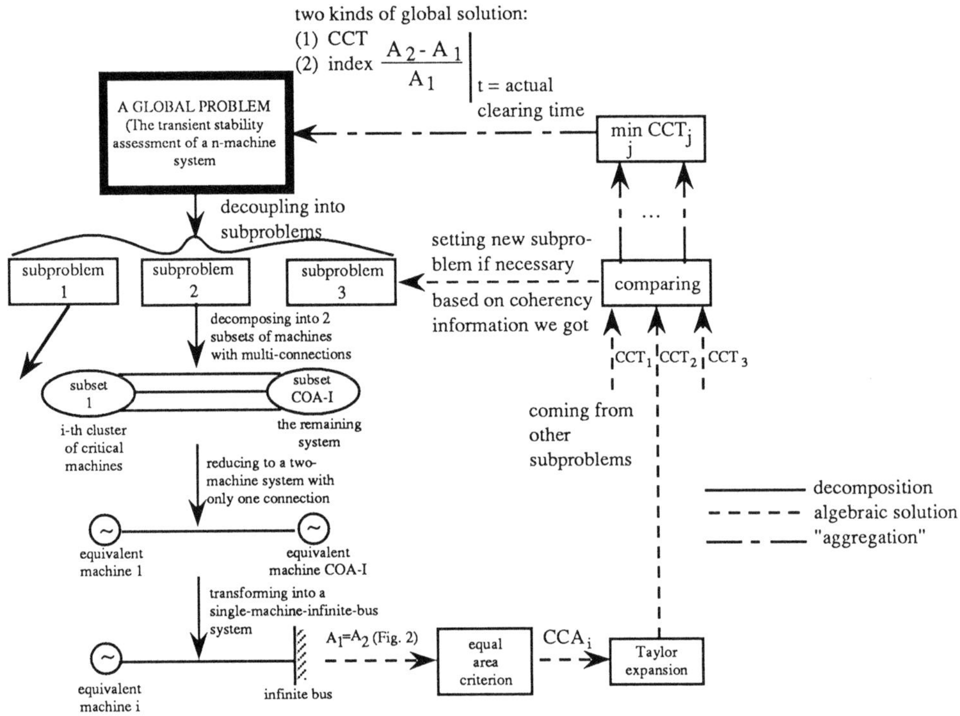

Figure 6.1 Principle of the EEAC method (Reproduced from Xue et al. [1987])

6.6 Dynamic Security Assessment

We begin the discussion of this topic by an observation by Findlay et al. [1988]. "The increasing pressure on minute by minute operations due to complex operating limits and reducing operating flexibility from spare capacity is driving many utilities to provide support in the form of short term operating plans. The objective of dynamic security assessment is to ensure that the system will survive any recognized contingency. On-line time domain simulation of all recognized contingencies, applied to the current operating condition is the only comprehensive way of ensuring the survival of the power system following such contingencies. There is therefore no practical solution because of the excessive amount of computations required and the repeated simulation required to obtain an assessment. The common practice in the industry is to perform off-line studies using a manually identified set of critical or pessimistic operating conditions. The results of these studies are provided to the operator as an instruction and/or in a computerized form for the on-line monitoring of the power system but will not encompass many situations which the operators may face. On-line computation of approximate limits is currently limited to the use of various forms of the equal area criterion. Advanced methods should be developed to provide both accuracy and speed before on-line DSA becomes a reality."

The above assessment by the industry at first sight might seem pessimistic. But the research in using energy functions over the past four decades offers hope. Considerable developmental activity has taken place in the last 10 years to warrent such optimism. It is now agreed that Lyapunov/Energy functions offer a quick method for screening contingencies while detailed simulations using parallel processing techniques may offer the speed and accuracy desired in the future.

In a recent survey of utilities in Kumar et al. [1988], a majority of them expressed concerns about stability related constraints. Dynamic security assessment is defined as the evaluation of the ability of a power system to withstand a defined set of contingencies and to survive the transition to an acceptable steady state condition. To do this the operator should be able to simulate the stability crisis for the particular disturbance in some manner and answer questions regarding final steady state configuration. It may be either stable or unstable. In the case of the former, the margin of stability for postulated contingencies in a dynamic sense is a desired quantity to monitor. In the event of possible instability, some sort of preventive action is called for. This is done on the basis of sensitivity of the energy margin to a given contingency.

Application of energy functions to DSA was first proposed by Fouad et al. [1982] for contingency ranking using the energy margin technique. Later on a sensitivity based approach, both analytical and numerical, was proposed by Sauer et al. [1983] and Pai et al. [1984] for computing limits on interface flows and system loadability. El-Kady et al. [1986] later demonstrated that numerical sensitivities can give useful information in practical large networks. Recently Xue et al. [1988] have applied the sensitivity approach based on a known mode of instability which essentially reduces the system to a two-machine equivalent.

6.6.1 Contingency analysis using energy margin (Fouad et al. [1982])

For a given initial operating condition and topology, disturbances at different locations can be simulated. For each disturbance, the transient energy function at the instant of fault clearing $V_{c\ell}$ and the critical energy V_{cr} depending on the post-disturbance network is computed. Let $\Delta V = V_{cr} - V_{c\ell}$. For the same operating conditions, various post-fault scenarios (including sequential breaker operation) are also simulated and ΔV computed for each. This energy margin is normalized as

$$\Delta V\Big|_n = \frac{\Delta V}{V_{KE}\Big|_{corr}} \tag{6.44}$$

Severity of faults are then ranked depending on the value of $\Delta V|_n$. Faults having a small $\Delta V|_n$ are more severe. Depending on these values the operator can be alerted as to the severity of the faults. For low values of $\Delta V|_n$ preventive action is suggested. In system planning this technique is useful as a screening aid. For numerical results on practical systems see the reference cited above.

6.6.2 DSA using sensitivity concepts

Sensitivity theory has been applied with good success in contingency analysis and static security assessment of power systems. Rich literature exists in sensitivity theory of nonlinear dynamical systems. Its application for power systems was first proposed by Teichgraber et al. [1969] and Venkata et al. [1973] for single machine infinite bus systems and later extended by Sauer et al. [1983] for multimachine power systems.

Sensitivity Theory and Motivation for Power System Applications

Consider a dynamical system

$$\dot{x}(t) = f(x(t), p(t), t) \tag{6.45}$$

$$x(t_o) = x_o$$

where p(t) is a parameter vector. Given p(t) it is easy to map from the subspace of parameter variations onto the subspace in state space (Fig. 6.2). D_p is the subspace of parameter variations around the nominal point p_o and D_x represents the corresponding subspace in state space. We would rather have

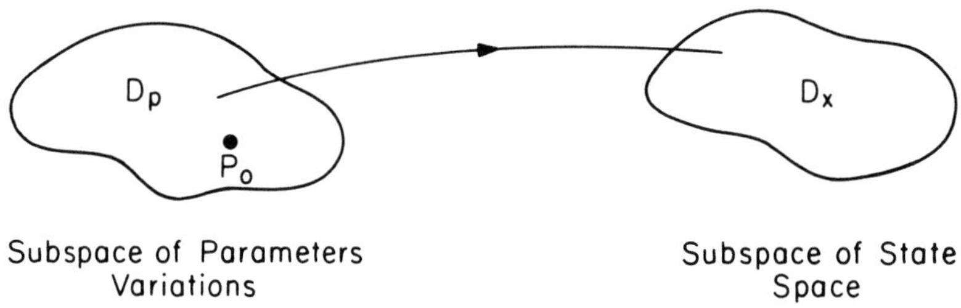

Figure 6.2 Sensitivity concept.

a relationship like

$$D_x = u(x,p_o t)D_p \tag{6.46}$$

where D_x is expressed as the product of a characteristic function $u(x,p_o,t)$ called the sensitivity function and a given set of parameter variations D_p. If we can precompute these sensitivity functions then for a given p_o and D_p we can compute D_x. In sensitivity theory $u(x,p_o,t)$ are obtained as solutions of differential equations which are solved along with (6.45). Details of the structure of these sensitivity equations are found in the books by Tomovic [1963] and Tomovic and Vukobratovic [1972].

Let us motivate the application of this to power systems by considering (i) a single machine infinite bus system and (ii) variation of a single parameter namely the pre-fault mechanical input. Referring to Fig. 1.4, which is self explanatory, the system is stable in $A_1 < A_2$ by equal area criterion. It is shown in the literature that this can be expressed in terms of the energy function as $A_1 + A_3 < A_2 + A_3$ which is equivalent to $V(\delta,\omega) < V_{cr}$. $V(\delta,\omega)$ is the energy function (or in this case the Lyapunov function) and V_{cr} is $V(\delta^u,0)$.

δ^u is the nearest unstable equilibrium point. In Teichgraeber et al. [1969] a stability measure ξ has been defined to be

$$\xi = 1 - \frac{V_{c\ell}}{V_{cr}} = \frac{V_{cr} - V_{c\ell}}{V_{cr}} \quad . \tag{6.47}$$

$V_{cr} - V_{c\ell}$ is defined to be the energy margin EM. One could alternatively define a quantity $\frac{A_2}{A_1}$ which relates the energy absorbing capacity of the post disturbance network to the kinetic energy at clearing. For multimachine systems this is termed as normalized potential energy margin with the modification that the kinetic energy is corrected to remove the K.E. associated with the center of inertia. In the case of single machine case ξ has been used by Teichgraeber et al. [1969] to show that $\frac{\partial \xi}{\partial P_m} < 0$ for a stable system and that $\left|\frac{\partial \xi}{\partial P_m}\right|$ can be used to provide useful information about how much more the prefault loading can be before the system becomes unstable for that fault (contingency). It can be seen that as P_m increases the energy margin decreases and so does ξ. Therefore $\frac{\partial \xi}{\partial P_m} < 0$. There is a limit to P_m beyond which the system will be unstable for the given fault and a given clearing time.

6.6.3 Multimachine Formulation - Analytical Expressions (Pai et al. [1984])

In a multimachine case the methodology with respect to a classical model is explained first and the parameter is chosen as the system loadability P_{TOT}. For on-line application it may be preferable to have analytical expressions for the sensitivities which are now discussed. In addition to the usual assumptions in the classical swing model of a multimachine system, two additional assumptions are made to facilitate an analytical development.

(1) Transfer conductances are neglected.

(2) Only self-clearing faults with no line switching are assumed so that the pre-fault network is equal to the post-fault network.

The differential equations during the faulted and the post-fault states are given in the standard center of angle notation as

$$M_i \ddot{\theta}_i = f^F_i(\theta) \quad , \quad 0 < t \leq t_{c\ell} \tag{6.48}$$

$$M_i \ddot{\theta}_i = f_i^{PF}(\theta) \qquad t > t_{c\ell} \tag{6.49}$$

The functional form of $f_i(\theta)$ with zero transfer conductances is

$$f_i(\theta) = P_{mi} - E_i^2 G_{ii} - \sum_{\substack{j=1 \\ \neq i}}^{n} C_{ij} \sin(\theta_i - \theta_j) \tag{6.50}$$

where $C_{ij} = E_i E_j B_{ij}$ and other symbols have the usual meaning. B_{ij} assumes different values for $f_i^F(\theta)$ and $f_i^{PF}(\theta)$.

The energy function for (6.49) has been shown to be (Chapter III, Sec. 2.3.2, with $Dij \equiv 0$)

$$V(\theta,\tilde{\omega}) = V_{KE}(\tilde{\omega}) + V_{PE}(\theta) \tag{6.51}$$

where
$$V_{KE}(\tilde{\omega}) = \frac{1}{2} \sum_{i=1}^{n} M_i \tilde{\omega}_i^2 \tag{6.52}$$

$$V_{PE}(\theta) = -\sum_{i=1}^{n} P_i \, (\theta_i - \theta_i^s) - \sum_{i=1}^{n-1} \sum_{j=i+1}^{n} C_{ij} [\cos(\theta_i - \theta_j) - \cos(\theta_i^s - \theta_j^s)]$$

where θ^s is the post-fault stable equilibrium point which, for the faults with no line switching is the same as the pre-fault stable equlibrium point. The energy margin EM is $V_{cr} - V_{c\ell}$. V_{cr} can be approximated either as $V(\theta^u, 0)$ where θ^u is the controlling u.e.p. or $V_{PE}(\theta^*)$ where θ^* is the crossing of the PEBS by the faulted trajectory. We take the latter since it is faster to compute on-line. Now

$$EM = -\sum_{i=1}^{n} P_i(\theta_i^* - \theta_i^{c\ell}) - \sum_{i=1}^{n-1} \sum_{j=i+1}^{n} C_{ij} \cos\left[(\theta_i^* - \theta_j^*) - (\theta_i^{c\ell} - \theta_j^{c\ell})\right]$$

$$- \frac{1}{2} \sum_{i=1} M_i (\tilde{\omega}_i^{c\ell})^2 \qquad (6.53)$$

where $P_i = P_{mi} - E_i^2 G_{ii}$. We can take $P_{mi}(i=1,2,\ldots,n)$ as the parameters and develop the sensitivity functions as $u_{ij}(t) = \dfrac{\partial\theta_i}{\partial P_{mj}}$ $(i=1,2,\ldots,n,\ j=1,2,\ldots,n)$ resulting in n^2 sensitivity differential equations. Alternatively, P_{TOT} can be taken as the parameter with the sensitivity function defined as $u_i = \dfrac{\partial\theta_i}{\partial P_{TOT}}$ $(i=1,2,\ldots,n)$ where $P_{TOT} = \Sigma P_{Gi}$. In steady state $P_{Gi} = P_{mi}$. This results in only n sensitivity differential equations. The latter appproach is outlined below.

The sensitivity of EM to P_{TOT} is given by

$$S_{EM} = \frac{\partial EM}{\partial P_{TOT}} = -\sum_{i=1}^{n} (\theta_i^* - \theta_i^{c\ell}) \frac{\partial P_{mi}}{\partial P_{TOT}} - \sum_{i=1}^{n} \left(\frac{\partial\theta_i^*}{P_{TOT}} - \frac{\partial\theta_i^{c\ell}}{\partial P_{TOT}}\right) P_i$$

$$- \sum_{i=1}^{n} M_i \tilde{\omega}_i^{c\ell} \frac{\partial\tilde{\omega}_i^{c\ell}}{\partial P_{TOT}} - \sum_{j=1}^{n-1} \sum_{j=i+1}^{n} C_{ij} \left[\sin\theta_{ij}^* \frac{\partial\theta_{ij}^*}{\partial P_{TOT}} - \sin\theta_{ij}^{c\ell} \frac{\partial\theta_{ij}^{c\ell}}{\partial P_{TOT}}\right]$$

$P_{TOT} = \Sigma P_{Gi}$ and P_{Gi} is computed by the economic dispatch criteria. We may assume $P_{Gi} = c_i P_{TOT} + d_i$. Hence $\dfrac{\partial P_{mi}}{\partial P_{TOT}} = c_i$ since in the pre-fault state $P_{mi} = P_{Gi}$. We can write S_{EM} functionally as

$$S_{EM} = S_{EM}(\theta^{cl}, \theta^*, \tilde{\omega}^{c\ell}, u^*, u^{c\ell}, \dot{u}^{c\ell}, P_G) \qquad (6.54)$$

where $\qquad \dot{u}_i^{c\ell} = \dfrac{\partial\tilde{\omega}_i^{c\ell}}{\partial P_{TOT}}.$

The dynamic sensitivity equations are obtained as (see Pai et al. [1984] for details).

$$M_i \ddot{u}_i = c_i - \sum_{\substack{j=1 \\ \neq i}}^{n} C_{ij} \cos \theta_{ij}(u_i - u_j) \qquad i=1,2,\ldots,n \tag{6.55}$$

Equations (6.48), (6.49), (6.54) and (6.55) then constitute the composite energy margin sensitivity model, (6.55) is a set of time varying differential equations and the initial conditions are $u_i = \dot{u}_i = 0$.

Example 6.1: Single Machine System

Assuming zero electrical output during fault, the faulted system equation is

$$M\ddot{\delta} + D\dot{\delta} = P_m \tag{6.56}$$

The post-fault system is given by

$$M\ddot{\delta} + D\dot{\delta} + P_e \sin \delta = P_m \tag{6.57}$$

The energy function is given by

$$V(\delta,\omega) = \frac{1}{2} M\omega^2 - P_m(\delta-\delta^s) - P_e \left[\cos \delta - \cos \delta^s\right] \tag{6.57}$$

and

$$V_{cr} = - P_m(\delta^u - \delta^s) - P_e (\cos \delta^u - \cos \delta^s) \tag{6.58}$$

The energy margin is given by EM = $V_{cr} - V_{c\ell}$. Since there is no economic dispatch involved here, the sensitivity equation with respect to $u = \frac{\partial \delta}{\partial P_m}$ is given by

$$M\ddot{u} + D\dot{u} = 1 \qquad 0 < t \leqslant t_{c\ell} \tag{6.59}$$

$$M\ddot{u} + D\dot{u} + (P_e \cos \delta)u = 1 \qquad t > t_{c\ell} \tag{6.60}$$

Initial conditions are $u = \dot{u} = 0$.

Assuming a certain $t_{c\ell}$, Equations (6.56) and (6.59) are integrated and the quantity $S_{EM} = \frac{\partial(EM)}{\partial P_m}$ is computed. In the case of a single machine case SEM turns out to be

$$S_{EM} = \delta + u(P_m - P_e \sin \delta) - M\omega\dot{u} - \delta^u. \tag{6.61}$$

S_{EM} can be computed as a function of time but its value at $t = t_{c\ell}$ will be of interest.

In Behera [1988] the methodology of Section 6.6.3 has been extended to a detailed model of the machine and validated on a 39 bus 10 machine system. The method has also been validated for practical systems by Vittal et al. [1989] using the classical model and controlling u.e.p. method. They have also included transfer conductances and line switching following a disturbance. Another application which is based on numerical sensitivity is discussed next.

6.6.4 DSA in practical networks using sensitivity concepts

This section describes the work by El-Kady et al. [1986] in applying sensitivity analysis to compute the maximum of interface power limits on a large network. The controlling u.e.p. method is used in computing V_{cr} instead of the PEBS method.

The transient energy margin ε_K for contingency k is defined as $\Delta V_k = V_{cr} - V_{c\ell}$. It is a complicated function of pre- and post-fault parameters, generation levels and fault conditions. In practice, operating criteria are set in terms of interface flows I_i (i=1, 2 ... L) on certain lines. The limiting values I_i^{lim} of these interface flows in the pre-fault state for a set of contingencies is what we seek to compute and display to the operator. The operator can compare these limiting values to the current monitored values of I_i. Using time simulation I_i^{lim} also can be computed by a trial and error procedure. Suppose we know (from off-line computations) the sensitivities of the energy margin with respect to these interface flows as $S_{ki} = \frac{\partial \Delta V_k}{\partial I_i}$. Then the algorithm does the following.

1. Compute the distribution factors $\alpha_{ig} = \frac{dI_i}{dP_g}$ on-line from the load flow Jacobian matrix.

2. Calculate from the measurements the interface flow, I_i, which is of particular interest. Also measure the generator powers P_g.

3. For a given contingency k, integrate the fault-on trajectory until clearing to find $V_{c\ell}$.

4. For the given contingency, compute V_{cr} and hence find $\Delta V_k = V_{cr} - V_{c\ell}$.

5. If $\Delta V_k < 0$ this implies that the system will be unstable for this contingency. Carry out some preventive control action and go to Step 1. Otherwise go to step 6.

6. Calculate or retrieve S_{ki}.

7. If $\Delta V_k > 0$, use S_{ki} to compute I_i limit as

$$I_i^{new} = I_i^{old} + S_{ki}\,(0 - \Delta V_k^{old}) \tag{6.62}$$

This I_i^{new} may be displayed to the operator as well as the prespecified limiting values of I_i.

8. Use α_{ig} to compute P_g^{new} from

$$P_g^{new} \simeq P_g^{old} + \frac{1}{\alpha_{ig}} (I_i^{new} - I_i^{old}) \tag{6.63}$$

If P_g is adjusted to the P_g^{new} values then the limit I_i^{new} can be met in the event of contingency.

9. Go through a pre-selected set of contingencies and display appropriate information to the operator. He can take preventive action if necessary.

6.6.5 DSA using sensitivity method with known mode of instability

In Section 6.5 we talked about stability assessment based on a two-machine equivalent of the system, i.e., the critical generators and the rest of the system. It was also shown that this can be converted to a single machine infinite bus (SMIB) system whose swing equation is of the form

$$M\ddot{\delta} = P_m - [P_c + P_{max} \sin(\delta - \nu)] \tag{6.64}$$

The equal area criterion can be applied to this system as explained in Chapter III. Xue et al. [1988] use this formulation to compute sensitivity of the energy margin (EM) to changes in pre-fault loading conditions for a given contingency and a given $t_{c\ell}$. In the case of the SMIB system (Fig. 1.4), EM = $\frac{A_2}{A_1}$ where A2 and A_1 are decelerating and accelerating areas, respectively. In Sec. 6.6.3 (Example 6.1) we used the methodology of sensitivity analysis which resulted in a differential equation for the sensitivity coefficient $u = \frac{\partial \delta}{\partial P_m}$. An alter nate method for the SMIB system would be to express EM in terms of the system parameters and δ only. Then we take $\frac{\partial EM}{\partial P_m}$ which will be a complex expression involving several partial derivatives including $\frac{\partial \delta}{\partial P_m}$. For a given operating condition, contingency and clearing time $t_{c\ell}$, $\frac{\partial EM}{\partial P_m}$ is computed at $t = t_{c\ell}$ using either the fault-on trajectory or a trajectory approximation. This sensitivity information can be used to find limiting value of P_m which will make EM = pre-specified value. Second order sensitivities can also be used. The method has been tested on a large system with satisfactory results by Xue et al. [1988].

Conclusion

In this chapter we have talked about the applications of the energy function method to practical systems and dynamic security assessment. At the present time these tools remain as demonstrative packages only. They have yet to be integrated in an EMS data base to establish consistency and reliability of the results so far as the operator is concerned.

Chapter 7

FUTURE RESEARCH ISSUES

In this chapter we briefly survey what has been accomplished in this field so far and indicate new areas of research for further investigation. The history of using direct methods in power system stability is a long one of nearly four decades following the equal area criterion by Kimbark. Although there was a relative lull in the 60's and early 70's, the work being limited only to the academia, the late 70's and the 80's have seen a big upsurge in terms of research publications. There have been the breakthrough concepts of the controlling u.e.p and the potential energy boundary surface. The two are not irreconcilable concepts as suspected earlier. The characterization of the region of stability boundary in the multi-dimensional space has helped the development of hybrid techniques which combine the best features of both the controlling u.e.p and the PEBS methods. As systems became larger, consistency regarding accurate determination of critical clearing time or energy margin were matters of concern. Here the concept of grouping of machines following a fault has been helpful. This concept can be used either in a multimachine context by correcting the K.E and identifying the grouping through normalized P.E margin or reduce the system to a single machine infinite bus system by assuming coherency in the two groups. Extensions to multi-area energy functions have been made. If coherency is not to be assumed then corrections to the potential energy portion of the energy function associated with centers of inertia (slow energy function) are needed. There has been widespread interest in using energy functions with structure preserving models, the primary goal being that of having a more accurate representation of loads. The network dynamics being fast are represented as algebraic equations. There has also been considerable work going on in incorporating more dynamics in the model beyond the classical model. This includes the synchronous machine and the excitation system dynamics. The stability boundary has been approximated locally by hyperplanes or hypersurfaces passing through the controlling u.e.p. Globally there has been a proper characterization of the stability boundary through the concept of manifolds.

When one looks at such impressive and voluminous work done in the last decade, there is naturally a degree of optimism that the method will be used very soon as application software in an energy management system (EMS). But the question remains as to which method does one select out of the numerous methods that are available. The answer is not an easy one. What is lacking is a comparative assessment of the various algorithms for judging them against the same criteria. Some of the algorithms are backed by physical intution and well tested on large systems, but they still fail in some instances. Other algorithms while theoretically sound lack application to real systems. Some of future research topics in this area are

1. Comparison and testing of different methods for computing critical clearing time and energy margins for a given $t_{c\ell}$. Among the contending candidate methods are
    1. Controlling u.e.p method.
    2. Potential energy boundary surface method.
    3. Two area or multi-area energy function method.
    4. Hybrid method which combines (1) and (2).
2. Extension of the hybrid method to the structure preserving model.
3. Inclusion of dynamic load representation in the direct methods.

4. Use of better trajectory approximations to avoid simulation altogether. The Taylor series method has been used successfully so far but in a multi-machine context with detailed modeling it may not yield good results.
5. Investigate if incorporation of more machine and excitation dynamics is really needed. Recently there has been some effort in representing excitation control in a parametric manner.
6. In the context of the multi-area energy function approach, a technique to identify on-line the coherent areas depending on a given contingency will be really helpful.
7. Potential energy correction in the case of two area or multi-area energy function through rigorous mathematical approach is needed.
8. In view of the growing interest in voltage collapse or voltage instability, an integrated approach to both transient and voltage stability using direct methods may be possible using structure preserving energy functions.
9. A rigorous theory regarding stability of differential-algebraic systems will benefit power system stability research.
10. Since the operator would like to have a tool similiar to that of a planner namely simulation, it may be worthwhile to pursue research in parallel processing as a way of fast simulation and use direct methods as a filter to select critical contingencies only. Methods based on wave-form relaxation technique which have been successful in VLSI circuits have been proposed and some simulation on parallel processors such as the Hypercube have been reported.
11. Application of the Artifical Intelligence approach or the knowledge-based expert systems together with direct methods and/or parallel processing may be a fruitful area of research.

APPENDIX A: 10 Machine System Data

The single line diagram, bus data, line, generator and excitation system data are given in the following pages. They are taken from Athay et al [1979].

TABLE A.1

BUS DATA

Bus	Type (0=P-Q) (1=P-V)	Volts	Load MW	Load MVAR	Gen Load	Gen MVAR
1	0	-	0.0	0.0	0.0	0.0
2	0	-	0.0	0.0	0.0	0.0
3	0	-	322.0	2.40	0.0	0.0
4	0	-	500.0	184.0	0.0	0.0
5	0	-	0.0	0.0	0.0	0.0
6	0	-	0.0	0.0	0.0	0.0
7	0	-	233.8	84.0	0.0	0.0
8	0	-	522.0	176.0	0.0	0.0
9	0	-	0.0	0.0	0.0	0.0
10	0	-	0.0	0.0	0.0	0.0
11	0	-	0.0	0.0	0.0	0.0
12	0	-	7.5	88.0	0.0	0.0
13	0	-	0.0	0.0	0.0	0.0
14	0	-	0.0	0.0	0.0	0.0
15	0	-	320.0	153.0	0.0	0.0
16	0	-	329.0	32.3	0.0	0.0
17	0	-	0.0	0.0	0.0	0.0
18	0	-	158.0	30.0	0.0	0.0
19	0	-	0.0	0.0	0.0	0.0
20	0	-	628.0	103.0	0.0	0.0
21	0	-	274.0	115.0	0.0	0.0
22	0	-	0.0	0.0	0.0	0.0
23	0	-	247.5	84.60	0.0	0.0
24	0	-	308.6	-92.20	0.0	0.0
25	0	-	224.0	47.20	0.0	1.0
26	0	-	139.0	17.0	0.0	0.0
27	0	-	281.0	75.5	0.0	0.0
28	0	-	206.0	27.6	0.0	0.0
29	0	-	283.5	26.9	0.0	0.0
30	1	1.0475	0.0	0.0	250.0	-
31	1	.982	9.2	4.60	-	-
32	1	.9831	0.0	0.0	650.0	-
33	1	.9972	0.0	0.0	632.0	-
34	1	1.0123	0.0	0.0	508.0	-
35	1	1.0493	0.0	0.0	650.0	-
36	1	1.0635	0.0	0.0	560.0	-
37	1	1.0278	0.0	0.0	540.0	-
38	1	1.0265	0.0	0.0	830.0	-
39	1	1.03	1104.0	250.0	1000.0	-

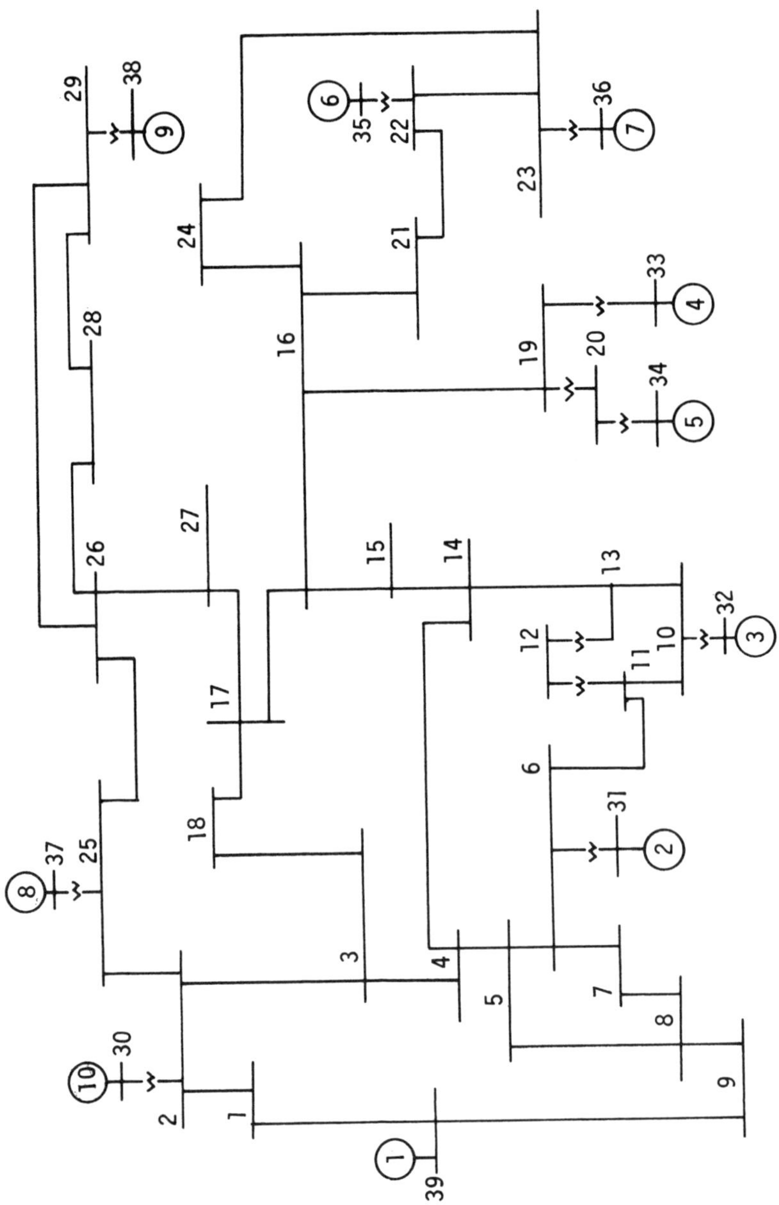

Fig. A.1 10 unit 39 bus New England test system.

TABLE A.2

LINE DATA FOR 10 UNIT SYSTEM

Line Data					Transformer Tap	
Bus	Bus	Resistance	Reactance	Susceptance	Magnitude	Angle
1	2	.0035	.0411	.6987	.000	.00
1	39	.0010	.0250	.7500	.000	.00
2	3	.0013	.0151	.2572	.000	.00
2	25	.0070	.0086	.1460	.000	.00
3	4	.0013	.0213	.2214	.000	.00
3	18	.0011	.0133	.2138	.000	.00
4	5	.0008	.0128	.1342	.000	.00
4	14	.0008	.0129	.1382	.000	.00
5	6	.0002	.0026	.0434	.000	.00
5	8	.0008	.0112	.1476	.000	.00
6	7	.0006	.0092	.1130	.000	.00
6	11	.0007	.0082	.1389	.000	.00
7	8	.0004	.0046	.0780	.000	.00
8	9	.0023	.0363	.3804	.000	.00
9	39	.0010	.0250	1.2000	.000	.00
10	11	.0004	.0043	.0729	.000	.00
10	13	.0004	.0043	.0729	.000	.00
13	14	.0009	.0101	.1723	.000	.00
14	15	.0018	.0217	.3660	.000	.00
15	16	.0009	.0094	.1710	.000	.00
16	17	.0007	.0089	.1342	.000	.00
16	19	.0016	.0195	.3040	.000	.00
16	21	.0008	.0135	.2548	.000	.00
16	24	.0003	.0059	.0680	.000	.00
17	18	.0007	.0082	.1319	.000	.00
17	27	.0013	.0173	.3216	.000	.00
21	22	.0008	.0140	.2565	.000	.00
22	23	.0006	.0096	.1846	.000	.00
23	24	.0022	.0350	.3610	.000	.00
25	26	.0032	.0323	.5130	.000	.00
26	27	.0014	.0147	.2396	.000	.00
26	28	.0043	.0474	.7802	.000	.00
26	29	.0057	.0625	1.0290	.000	.00
28	29	.0014	.0151	.2490	.000	.00
12	11	.0016	.0435	.0000	1.006	.00
12	13	.0016	.0435	.0000	1.006	.00
6	31	.0000	.0250	.0000	1.070	.00
10	32	.0000	.0200	.0000	1.070	.00
19	33	.0007	.0142	.0000	1.070	.00
20	34	.0009	.0180	.0000	1.009	.00
22	35	.0000	.0143	.0000	1.025	.00
23	36	.0005	.0272	.0000	1.000	.00
25	37	.0006	.0232	.0000	1.025	.00
2	30	.0000	.0181	.0000	1.025	.00
29	38	.0008	.0156	.0000	1.025	.00
19	20	.0007	.0138	.0000	1.060	.00

TABLE A.3

DETAILED MODEL UNIT DATA 10 UNIT SYSTEM

Unit No.	H (sec)	R_a	x'_d	x'_q	x_d	x_q	T'_{do}	T'_{qo}	x_ℓ
1	500.0	0	.006	.008	.02	.019	7.0	.7	.003
2	30.3	0	.0697	.170	.295	.282	6.56	1.5	.035
3	35.8	0	.0531	.0876	.2495	.237	5.7	1.5	.0304
4	28.6	0	.0436	.166	.262	.258	5.69	1.5	.0295
5	26.0	0	.132	.166	.67	.62	5.4	.44	.054
6	34.8	0	.05	.0814	.254	.241	7.3	.4	.0224
7	26.4	0	.049	.186	.295	.292	5.66	1.5	.0322
8	24.3	0	.057	.0911	.290	.280	6.7	.41	.028
9	34.5	0	.057	.0587	.2106	.205	4.79	1.96	.0298
10	42.0	0	.031	.008	.1	.069	10.2	.0	.0125

TABLE A.4

DETAILED MODEL UNIT EXCITATION SYSTEM DATA 10 UNIT SYSTEM

Unit No.	K_A	T_A	V_{RMIN}	V_{RMAX}	K_E	T_E	K_F	T_F	C1	C2
1	0	0	0	0	0	0	0	0	0	0
2	6.2	.05	-1.0	1.0	-.633	.405	.057	.5	.66	.88
3	5.0	.06	-1.0	1.0	-.0198	.5	.08	1.0	.13	.34
4	5.0	.06	-1.0	1.0	-.0525	.5	.08	1.0	.08	.314
5	40.0	.02	-10.0	10.0	1.0	.785	.03	1.0	.07	.91
6	5.0	.02	-1.0	1.0	-.0419	.471	.0754	1.246	.064	.251
7	40.0	.02	-6.5	6.5	1.0	.73	.03	1.0	.53	.74
8	5.0	.02	-1.0	1.0	-.047	.528	.0854	1.26	.072	.282
9	40.0	.02	-10.5	10.5	1.0	1.4	.03	1.0	.62	.85
10	5.0	.06	-1.0	1.0	-.0485	.25	.04	1	.08	.26

NOTE: A & B in Figure A.2 are computed as follows:

$$EX_2 = \frac{V_{RMAX}}{K_E + C2} \quad , \quad EX_1 = .75\ EX_2$$

$$B = \ell n(C2/C1)/(EX_2 - EX_1) \quad , \quad A = C2/e^{B \cdot EX_2}$$

Unit #1 has constant excitation.

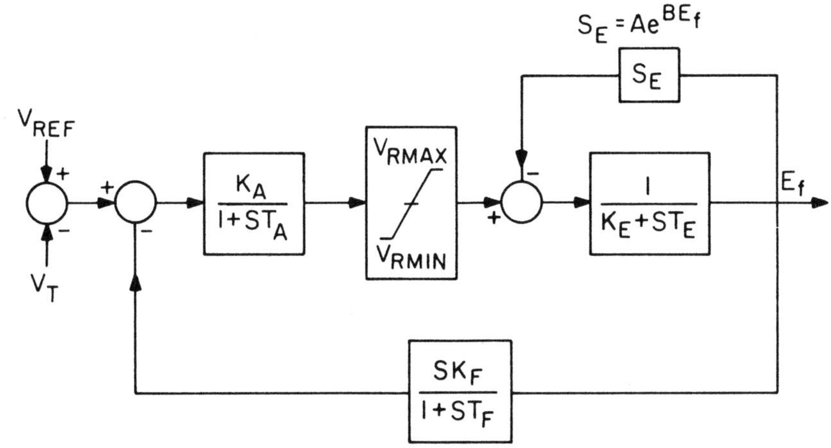

Figure A.2 - IEEE Type 1 rotating excitation system model

REFERENCES

Anderson, B. D. O., "A system theory criterion for positive real matrices," SIAM J. Control, vol. 5, no. 2, 1967, pp. 171-182.

Athay, T., R. Podmore and S. Virmani, "A practical method for direct analysis of transient stability," IEEE Trans. on Power Apparatus and Systems, vol. PAS-98, no. 2, March/April 1979, pp. 573-584.

Athay, T., V. R. Sherket, R. Podmore, S. Virmani and C. Puech, Conference on System Engineering for Power: Emergency Operating State Control - Section IV, Davos, Switzerland, Sept. 1979. Also U.S. Dept. of Energy Publication No. CONF-790904-PL, 1980.

Athay, T., and D. I. Sun, "An inproved energy function for transient stability analysis," Proc. International Symposium on Circuits and Systems, Chicago, April 1981.

Aylett, P. D., "The energy-integral criterion of transient stability limits of power systems," Proc. of the Institution of Electrical Engineers (London), vol. 105C, no. 8, Sept. 1958, pp. 257-536.

Behera, A. K., "Transient stability analysis of multi-machine power systems using detailed machine models," Ph.D. Thesis, University of Illinois at Urbana-Champaign, 1988.

Bergen, A. R., and J. Hill, "Structure preserving model for power system stability analysis," IEEE Trans. Power Appar. Syst., vol. PAS-100, no. 1, Jan. 1981.

Bergen, A. R., D. J. Hill and C. L. DeMarco, "A Lyapunov function for multimachine power systems with generator flux decay and voltage dependent loads," International Journal of Electric Power and Energy Systems, vol. 8, no. 1, Jan. 1986, pp. 2-10.

Carvalho, V. F., M. A. El-Kady, E. Vaahedi, P. Kundur, C. K. Tang, G. Rogers, J. Libaque, D. Wong, A. A. Fouad, V. Vittal and S. Rajagopal, "Direct analysis of transient stability for large power systems," Report EL-4980 Project RP 2206-1, EPRI, Palo Alto, Dec. 1986.

Chandrashekhar, K. S., and D. J. Hill, "Dynamic security dispatch: basic formulation," IEEE Trans. on Power Apparatus and Systems, vol. PAS-102, no. 7, July 1983, pp. 2145-2154.

Chandrashekhar, K. S., and D. J. Hill, "Cutset stability criterion for power systems using a structure preserving model," Int. Journal of Electric Power and Energy Systems, vol. 8, no. 3, July 1986, pp. 146-157.

Chiang, H. D., F. F. Wu and P. P. Varaiya, "Foundations of direct methods for power system transient stability analysis," IEEE Trans. Circuits and Systems, vol. CAS-34, Feb. 1987, pp. 160-173.

Chiang, H. D., F. F. Wu and P. P. Varaiya, "Foundations of the potential energy boundary surface method for power system transient stability analysis," IEEE Trans. on Circuits and Systems, vol. CAS-35, June 1988, pp. 160-172.

Chiang, H. D., M. W. Hirsch and F. F. Wu, "Stability regions of nonlinear autononomous dynamical systems," IEEE Transactions on Automatic Control, vol. 33, no. 1, Jan. 1988, pp. 16-27.

Chong, C. N., "Energy functions for nonlinear structure preserving models," Ph.D. Thesis, University of New Castle, Australia, Aug. 1986.

Chow, J. H., J. J. Allemong and P. V. Kokotovic, "Singular perturbation analysis of systems with sustained high frequency oscillations," Automatica, vol. 14, 1978, pp. 271-279.

Chow, J. H., Time Scale Modeling of Dynamic Networks with Application to Power Systems, Springer-Verlag, New York, 1982.

Cook, P. A., and A. M. Eskicioglu, "Transient stability analysis of power systems by the method of tangent hypersurfaces," Proc. IEE, vol. 130, Part C, no. 4, July 1983, pp. 183-192.

CIGRE Committee 32 Report, C. Barbier, L. Carpentier and F. Saccomanno, "Tentative classification and terminologies relating to stability problems in power systems," Electra, no. 56, 1978.

Desoer, C. A., and M. Y Wu, "Stability of a nonlinear time invariant feedback system under almost constant inputs," Automatica, vol. 5, 1969, pp. 231-233.

El-Abiad, A. H., and K. Nagappan, "Transient stability region of multi-machine power sytems," IEEE Trans. on Power Apparatus and Systems, vol. PAS-85, no. 2, Feb. 1966, pp. 169-178.

El-Kady, M. A., C. K. Tang, V. F. Carvalho, A. A. Fouad and V. Vittal, "Dynamic security assessment utilizing the transient energy function method," IEEE Trans. on Power Systems, vol. PWRS1, no. 3, Aug. 1986, pp. 284-291.

Findlay, J. A., V. F. Carvalho, G. A. Maria and C. E. Graham, "State of the art and key issues in power system security assessment," Proceedings of the Workshop on Power System Security Assessment, April 27-29, 1988, Ames, IA.

Fouad, A. A., "Stability theory-criteria for transient stability," Proc. Engineering Foundation Conference on "System Engineering for Power," Henniker, New Hamsphire, 1975, pp. 421-450.

Fouad, A. A., S. E. Stanton, K. R. C. Mamandur and K. C. Kruempel, "Contingency analysis using transient energy margin technique," IEEE Transactions on Power Apparatus and Systems, vol. PAS-101, no. 4, April 1982, pp. 757-766.

Fouad, A. A., and V. Vittal, "Power system response to a large disturbance: energy associated with system separation," IEEE Trans. on Power Apparatus and Systems, vol. 102, Nov. 1983, pp. 3534-3540.

Fouad, A. A., V. Vittal, S. Rajagopal, V. F. Carvalho, M. A. El-Kady, C. K. Tang, J. V. Mitsche and M. V. Pereira, "Direct transient stability analysis using energy functions: application to large power networks," IEEE Trans. on Power Systems, vol. PWRS-2, Feb. 1987, pp. 37-44.

Fouad, A. A., V. Vittal, Y. X. Ni, A. R. Pota, K. Nodehi and T. K. Oh, "Extending applications of the transient energy function method," Report EPRI EL-5215, Sept. 1987, EPRI, Palo Alto, CA.

Fouad, A. A., V. Vittal, Y-X. Ni, H. M. Zein-Eldin, E. Vaahedi, H. R. Pota, K. Nodehi and J. Kim, "Direct transient stability assessment with excitation control," Paper 88 SM 657-9, IEEE Power Engineering Society Summer Meeting, Portland, July 24-29, 1988.

Ghosh, K. K., "Direct stability evaluation and on-line dynamic security assessment of power systems using structure preserving energy functions," Ph.D. Thesis, I.I.T. Kanpur, Feb. 1987.

Gill, P. E., and W. Murray, "Algorithms for the solution of nonlinear least squares problems," Amer. J. Numer. Anal., vol. 5, no. 5, Oct. 1978, pp. 977-992.

Gless, G. E., "Direct method of Lyapunov applied to transient power system stability," IEEE Trans. on Power Apparatus and Systems, vol. PAS-85, Feb. 1966, pp. 164-179.

Hill, D. J., and C. N. Chong, "Lyapunov functions of Lure'-Postninkov form for structure preserving models of power systems," Tech. Report EE 8529, University of New Castle, Australia, Aug. 1985.

IEEE Task Force on terms and definitions "Proposed terms and definitions for power system stability," IEEE Trans. on Power Apparatus and Systems, vol. PAS-101, no. 7, July 1982, pp. 1894-1898.

Kakimoto, N., Y. Ohsawa and M. Hayashi, "Transient stability analysis of electric power system via Lure' type Lyapunov functions, parts I and II," Trans. IEE of Japan, vol. 98, no. 516, May/June 1978.

Kakimoto, N., Y. Ohsawa and M. Hayashi, "Construction of Lure' type Lyapunov function with effect of magnetic flux decay," Memoirs of the Faculty of Engineering, Kyoto University, vol. XLI, Part 2, April 1979, Kyoto, Japan.

Kakimoto, N., Y. Ohsawa and M. Hayashi, "Transient stability analysis of multi-machine power systems with field flux decay via Lyapunov's direct method," IEEE Trans. Power Appar. Syst., vol. PAS-99, no. 5, Sept.-Oct. 1980, pp. 1819-1827.

Kakimoto, N., Y. Ohnogi. H. Matsuda and H. Shibuya, "Transient stability analysis of large scale power systems by Lyapunov's direct method," IEEE Trans. on Power Apparatus and Systems, vol. PAS-103, no. 1, Jan. 1986, pp. 160-167.

Kimbark, E. W., Elements of Stability Calculations, vol. 1, John Wiley and Sons, New York, 1948.

Kokotovic, P. V., B. Avramovic, J. H. Chow and J. R. Winkelman, "Coherency-based decomposition and aggregation," Automatica, vol. 18, 1982, pp. 47-56.

Kumar, A. B. R., H. Amelink, C. K. Pang, M. A. Pai and P. W. Sauer, "Dynamic security assessment for power systems: research plan," Report EL-4958, Project 2496-1, EPRI, Palo Alto, CA, Aug. 1987.

LaSalle, J. P., and S. Lefschetz, "Stability by Lyapunov's direct method with applications," Academic Press, New York, 1961.

Lemmon, W., K. C. Mamandur and W. R. Barcelo, "Transient stability prediction and control in real time by QUEP," Paper 88 SM 671-0, IEEE PES Summer Power Meeting, Portland, July 24-29, 1988.

Magnusson, P. C., "Transient energy method of calculating stability," AIEE Trans., vol. 66, 1947, pp. 747-755.

Marino, R., and S. Nicosia, "Hamiltonian-type Lyapunov functions," IEEE Transactions on Automatic Control, vol. AC-28, no. 11, Nov. 1983, pp. 1055-57.

Michel, A. N., N. R. Sarabudla and R. K. Miller, "Stability analysis of complex dynamical systems with computational methods," Circuits, Signals and Signal Processing, vol. 1, 1982, pp. 171-202.

Michel, A. N., A. A. Fouad and V. Vittal, "Power system transient stability using individual machine energy functions," IEEE Trans. Circuits Syst., CAS-30, no. 5, May 1983, pp. 266-276.

Miller, R. K. and A. N. Michel, Ordinary Differential Equations, Academic Press, New York, 1982.

Miyagi, H., and T. Taniguchi, "Construction of Lyapunov functions for power systems," Proc. IEE, vol. 124, no. 12, 1977.

Miyagi, H., and A. R. Bergen, "Stability studies of multimachine power systems with effects of automatic voltage regulators," IEEE Trans. Automatic Control, vol. AC-31, no. 3, March 1986, pp. 210-215.

Narasimhamurthi, N., and M. R. Musavi, "A general energy function for transient stability analysis of power systems," IEEE Trans. Circuits and Systems, vol. CAS-31, no. 7, July 1984, pp. 637-645.

Ni, Y. X. and A. A. Fouad, "A simplified two terminal HVDC model and its use in direct transient stability assessment," IEEE Transactions on Power Systems, vol. PWRS-2, no. 4, Nov. 1987, pp. 1006-1013.

Nityanandan, B. N., and K. Nagappan, "A new approach to transient stability of large scale power systems," Proc. IFAC Symposium on "Computer Applications in Large Scale Power Systems," New Delhi, India, Aug. 1979.

Nityanandan, B. N., "Decomposition and aggregation of multimachine power systems for transient stability," Ph.D. Thesis, Madurai University, Madurai, India, Aug. 1980.

Othman, H. A., "Decoupled stability analysis of power systems with slow and fast dynamics, Ph.D. Thesis, University of Illinois, Urbana-Champaign, Dec. 1987.

Othman, H. A., and M. A. Pai, "Dynamic equivalencing and direct methods of stability analysis," Proc. IFAC Symposium on "Power System Modeling and Control Applications," Brussels, Belgium, Sept. 5-8, 1988.

Padiyar, K. R., and H. S. Y. Sastry, "Topological energy function analysis of stability of power systems," International Journal of Electric Energy and Power Systems, vol. 9, no. 1, Jan. 1987, pp. 9-16.

Padiyar, K. R., and K. K. Ghosh, "A novel structure preserving energy function for direct stability evaluation of power systems with known modes of stability," Electric Machines and Power Systems, vol. 13, no. 2, 1987.

Padiyar, K. R., and K. K. Ghosh, "Direct stability evaluation of power systems with detailed generator models using structure preserving energy functions," International Journal of Electric Power and Energy Systems, vol. 11, no. 1, Jan. 1989, pp. 47-56.

Pai, M. A., "Power system stability studies by Lyapunov-Popov approach," Proc. 5th IFAC World Congress, Paris, 1972.

Pai, M. A., and V. Rai, "Lyapunov-Popov stability analysis of synchronous machine with flux decay and voltage regulator," Int. Journal of Control, 1974, vol. 19, no. 4, pp. 817-829.

Pai, M. A., Power System Stability, North Holland Publishing Co., New York, 1981.

Pai, M. A., P. W. Sauer and K. D. Demaree, "Direct methods of stability analysis in dynamic security assessment, Proc. 9th IFAC World Congress, Budapest, July 1984.

Pai, M. A., "Survey of practical direct methods of stability analysis in power systems, Electric Machines and Power Systems, vol. 9, March-June 1984, pp. 131-144.

Pai, M. A., H. Othman, J. H. Chow and J. R. Winkelman, "Time scale energies in large power systems," Proc. IFAC Symposium on Large Scale Systems, Zurich, Aug. 1986.

Racz, L. Z., and B. Bokay, Power System Stability, Elsevier Science Publishers, New York, 1988.

Rahimi, A., and G. Schaffer, "Power system transient stability indices of "worst case" dynamic contingencies," IEEE Transactions on Power Systems, vol. PWRS, no. 2, Aug. 1987, pp. 660-668.

Rai, V., "Transient stability analysis of power sytem using Lyapunov functions with improved synchronous machine models," Ph.D. Thesis, October 1973, I.I.T., Kanpur, India.

Rau, V. G., and N. Krishnamoorthy, "Operational transform technique for generating Lyapunov functions," Int. Journal of Electric Energy and Power Systems, vol. 6, no. 1, Jan. 1984, pp. 24-30.

Ribbens-Pavella, M., "On-line measurements of transient stability index," in Computerized Operation of Power Systems, C. S. Savulescu (Editor), Elsevier Scientific Publishing Co., New York, 1976.

Ribbens-Pavella, M., Lj. Gruije, J. Sabatel and A. Bouffioux, "Direct methods for stability analysis of large scale power systems," Proc. of IFAC Symposium on "Computer Applications in Large Scale Power Systems," vol. II, pp. 168-175, Aug. 1979, New Delhi, India.

Ribbens-Pavella, M., P. G. Murthy and J. L. Howard, "The acceleration approach to practical stability domain estimation in power systems," Proc. of 20th IEEE Conference on Decision and Control, San Diego, Dec. 16-18, 1981, vol. I, pp. 471-476.

Ribbens-Pavella, M., P. G. Murthy, J. L. Howard and J. L. Carpentier, "Transient stability index for on-line stability assessment and contingency evaluation," Int. Journal of Power and Energy Systems, vol. 4, no. 2, April 1982.

Ribbens-Pavella, M., and F. J. Evans, "Direct methods for studying of the dynamics of large scale electric power systems-a survey," Automatica, vol. 21, no. 1, 1985, pp. 1-21.

Sasaki, H. "An approximate incorporation of field flux decay into transient stability analysis of multi-machine power systems by the second method of Lyapunov," IEEE Transcations on Power Apparatus and Systems, vol. PAS-98, no. 2, March/April 1979, pp. 473-483.

Sastry, H. S. Y., "Application of topological energy functions for the direct stability evaluation of power systems," I.I.T. Kanpur India, 1984.

Sauer, P. W., K. D. Demaree and M. A. Pai, "Stability limited load supply and interchange capability," IEEE Transactions on Power Apparatus and Systems, vol. PAS-102, Nov. 1983, pp. 3637-3643.

Sauer, P. W., "Reduced order dynamic modeling of machines and power systems," Proc. American Power Conference, Chicago, April 27-29, 1987.

Sauer, P. W., A. K. Behera, M. A. Pai, J. R. Winkelman and J. H. Chow, "Trajectory approximations for direct energy methods that use sustained faults with detailed power system models," PES Paper 88 SM 688-4, IEEE PES Summer Power Meeting, Portland, July 1988.

Sauer, P. W., A. K. Behera, M. A. Pai, J. R. Winkelman and J. H. Chow, "A direct method for transient stability analysis of power systems with detailed models," Electric Machines and Power Systems, vol. 15, no. 1, 1988, pp. 1-16.

Siddiqee, M. W., "Transient stability of an ac generator by Lyapunov's direct method," Int. Journal of Control, vol. 8, no. 2, 1968, pp. 131-144.

Siddiqee, M. W., and J. Peschon, "Application of equal-area criterion to multimachine power system stability problems," Paper 69 CP 120-PWR, IEEE Winter Power Meeting, New York, Jan. 26-31, 1969.

Tavora, C. J., and O. J. M. Smith, "Stability analysis of power systems," IEEE Trans. on Power Apparatus and Systems, PAS 91, no. 3, May/June 1971, pp. 1138-1144.

Tavora, C. J., and O. J. M. Smith, "Equilibrium analysis of power systems," IEEE Trans. on Power Apparatus and Systems, vol. PAS-91, no. 3, 1972, pp. 1131-1137.

Teichgraeber, F. W. Harris and G. L. Johnson, "New stability measure for multimachine power systems," IEEE Transactions on Power Apparatus and Systems, vol. PAS-89, no. 2, Feb. 1970, pp. 233-239.

Tomovic, R., Sensitivity Analysis of Dynamical Systems, McGraw Hill, New York, 1963.

Tomovic, R., and M. Vukobratovic, General Sensitivity Theory, American Elsevier Publishing Co., New York, 1972.

Thomas, R. J., and J. S. Thorp, "Towards a direct test for large scale electric power instabilities," Proc. 24th Conference on Decision and Control, Ft. Lauderdale, FL, Dec. 1985, vol. I, pp. 65-69.

Tsolas, N., A. Araposthasis and P. Varaiya, "A structure preserving energy function for power system transient stability analysis," IEEE Trans. on Circuits and Systems, vol. CAS 32, no. 10, Oct. 1985, pp. 1041-1049.

Van Cutsem, Th., and M. Ribbens-Pavella, "Structure preservin direct methods for transient stability analysis of power systems," Proc. 24th IEEE Conference on Decison and Control, Ft. Lauderdale, FL, Dec. 1985, pp. 70-76.

Varaiya, P. P., F. F. Wu and R. L. Chen, "Direct methods for transient stability analysis of power systems," Proc. IEEE, vol. 73, Dec. 1985, pp. 1703-1715.

Venkata, S. S., W. J. Eccles and J. H. Noland, "Multiparameter sensitivity analysis of power system stability by Popov's method," International Journal of Control, vol. 73, no. 2, 1973, pp. 291-304.

Vittal, V., "Power system stability using critical energy of individual machines," Ph.D. Thesis, Iowa State University, Ames, Iowa, 1982.

Vittal, V., S. Rajagopal, A. A. Fouad, M. A. El-Kady, E. Vaahedi and V. F. Carvalho, "Transient stability analysis of stressed power systems using the energy function method," IEEE Transactions on Power Systems, vol. 3, no. 1, Feb. 1988, pp. 239-244.

Vittal, V., E. Z. Zhou, C. Hwang and A. A. Fouad, "Derivation of stability limits using analytical sensitivity of the energy margin," Paper 89 WM 207-2, IEEE PES Winter Power Meeting, New York, Jan. 1989.

Willems, J., "Comments on transient stability of a.c. generator by Lyapunov's direct method," International Journal of Control, vol. 10, no. 1, 1969, pp. 113-116.

Willems, J., "Direct methods for transient stability studies in power system analysis," IEEE Trans. Automat Contr., vol. AC-16, no. 4, July-Aug. 1971, pp. 1469-81.

Xue, Y., Th. Van Cutsem and M. Ribbens-Pavella, "A simple direct method for transient stability assessment of large power systems," IEEE Trans. on Power Systems, vol. 3, no. 1, May 1988, pp. 400-412.

Xue, Y., Th. Van Cutsem and M. Ribbens-Pavella, "Extended equal area criterion: justifications, generalizations and applications," IEEE Power Engineering Society Winter Power Meeting, Feb. 1988, Paper 88 WM 217-2.

Xue, Y., Th. Van Cutsem and M. Ribbens-Pavella, "Real-time analytic sensitivity method for transient security assessment and preventive control," Proc. IEE, vol. 135, pt. C, no. 2, March 1988, pp. 107-117.

Yee, H., "Region of transient stability in state space for synchronous generator," Proc. IEE, vol. 122, no. 7, July 1975.

Yee, H., and B. D. Spalding, "Transient stability analysis of multi-machine power systems by the method of hyperplanes," IEEE Transactions on Power Apparatus and Systems, vol. PAS-96, no. 1, Jan/Feb 1977.

Yee, H., and M. J. Muir, "Transient stability of multimachine systems with saturable exciters," Proc. IEE, vol. 127, Part C, no. 1, Jan. 1980.

Zaborszky, J., G. Huang, B. Zheng and T. C. Leung, "On the phase portrait of a class of large nonlinear dynamic systems such as the power systems," IEEE Trans. on Automatic Control, vol. 33, no. 1, Jan. 1988, pp. 4-15.

INDEX

L

M

N

P

Q

R

S

T

U

V

Belkhayat, Mohamed Code 2714
5005164